Now you can learn
the abstract concepts of algebra
through concrete modeling!

Computer Interactive Algeblocks, Volumes 1 and 2

by Anita Johnston, Jackson Community College

Taking algebra from frustrating to fun!

Learning abstract concepts from a lecture or a book is one of the more difficult hurdles many students face when they enter the algebra curriculum. Frustration may lead some to give up before ever building the strong algebraic foundation necessary for them to move on to higher-level courses. Now, with *Computer Interactive Algeblocks*, you actually learn by doing *and* seeing algebra—giving you the skills you need to continue with confidence.

Three easy steps to mathematical success!

1. *Follow the examples:* On-screen examples are automatic and guide you every step of the way. Each example begins with an explanation of how to solve the problem at hand, and then proceeds with an animated demonstration of how to use *Algeblocks* to solve the problem.

2. *Practice the concepts:* Once comfortable with the examples, you're ready to perform practice exercises. The program gives you the option to solve equations with or without the help of *Algeblocks.*

3. *Self-test:* The self-test allows you to review the concepts presented in the *Algeblocks* lab without the guidance of the program. If you have any difficulty, simply click on "Examples" or "Practice" to review, then return to the self-test when you're ready.

System Requirements

Macintosh®**:** System 7.0 or higher, 68030 Processor or higher, 8MB RAM, color monitor.
Windows®**:** Windows 3.1 or higher, VGA, 386/40 MHz processor, 8MB RAM, 640 x 480 screen resolution.

To order a copy of *Computer Interactive Algeblocks, Volume 1* or *2*, please contact your college store or fill out the form on the back and return with your payment to Brooks/Cole.

ORDER FORM

_____Yes! Send me a copy of *Computer Interactive Algeblocks, Volume 1*
(Macintosh® ISBN: 0-534-95144-9)

_____Yes! Send me a copy of *Computer Interactive Algeblocks, Volume 1*
(Windows® ISBN: 0-534-95146-5)

_____Yes! Send me a copy of *Computer Interactive Algeblocks, Volume 2*
(Macintosh ISBN: 0-534-95443-X)

_____Yes! Send me a copy of *Computer Interactive Algeblocks, Volume 2*
(Windows ISBN: 0-534-95443-X)

_____Copies x $20.95* = _____

Residents of: AL, AZ, CA, CT, CO, FL, GA, IL, IN, KS, KY, LA, MA, MD, MI, MN, MO, NC, NJ, NY, OH, PA, RI, SC, TN, TX, UT, VA, WA, WI must add appropriate state sales tax.

Subtotal _____
Tax _____
Handling __$4.00__
Total Due _____

Payment Options

_____ Check or money order enclosed

Bill my _____VISA _____MasterCard _____American Express

Card Number: _____ Expiration Date: _____

Signature: _____

Note: Credit card billing and shipping addresses must be the same.

Please ship my order to: (Please print.)

Name _____

Institution _____

Street Address_____

City _____ State _____ Zip+4_____

Telephone ( )_____ e-mail _____

Your credit card will not be billed until your order is shipped. Prices subject to change without notice. We will refund payment for unshipped out-of-stock titles after 120 days and for not-yet-published titles after 180 days unless an earlier date is requested in writing from you.

Mail to:

Brooks/Cole Publishing Company
Source Code 8BCMA188
511 Forest Lodge Road
Pacific Grove, California 93950-5098
Phone: (408) 373-0728; Fax: (408) 375-6414
e-mail: info@brookscole.com

10/97

Student Solutions Manual
for Johnson/Mowry's
MATHEMATICS
A Practical Odyssey
3rd Edition

DEANN CHRISTIANSON
ELAINE WERNER

Brooks/Cole Publishing Company

I(T)P® *An International Thomson Publishing Company*

Pacific Grove • Albany • Belmont • Bonn • Boston • Cincinnati • Detroit
Johannesburg • London • Madrid • Melbourne • Mexico City • New York
Paris • Singapore • Tokyo • Toronto • Washington

Sponsoring Editor: *Linda Row*
Marketing Representative: *Jennifer Huber*
Editorial Assistant: *Melissa Duge*
Production: *Dorothy Bell*

Cover Design: *Roy R. Neuhaus*
Cover Illustration: *Amy L. Wasserman*
Printing and Binding: *Webcom Limited*

For more information, contact:

BROOKS/COLE PUBLISHING COMPANY
511 Forest Lodge Road
Pacific Grove, CA 93950
USA

International Thomson Editores
Seneca 53
Col. Polanco
11560 México, D. F., México

International Thomson Publishing Europe
Berkshire House 168-173
High Holborn
London WC1V 7AA
England

International Thomson Publishing GmbH
Königswinterer Strasse 418
53227 Bonn
Germany

Thomas Nelson Australia
102 Dodds Street
South Melbourne, 3205
Victoria, Australia

International Thomson Publishing Asia
60 Albert Street
#15-01 Albert Complex
Singapore 189969

Nelson Canada
1120 Birchmount Road
Scarborough, Ontario
Canada M1K 5G4

International Thomson Publishing Japan
Hirakawacho Kyowa Building, 3F
2-2-1 Hirakawacho
Chiyoda-ku, Tokyo 102
Japan

Printed in Canada

10   9   8   7   6   5   4   3

ISBN 0-534-35077-1

PREFACE

This *Student Solutions Manual* has been prepared to accompany the textbook, *Mathematics: A Practical Odyssey*, Third Edition, by David B. Johnson and Thomas A. Mowry. Detailed solutions for every other odd problem in the textbook are provided except for answers to historical questions, essay questions and some calculator exercises.

Feedback concerning errors, solution correctness and manual style would be appreciated. These and any other comments can be sent directly to us at the address below or in care of the publisher.

The authors would like to thank David B. Johnson and Thomas A. Mowry for the opportunity to participate in their project. We would also like to thank Assistant Editor, Linda Row, for her guidance and support. We hope that students and instructors will find this manual to be a useful instructional aid.

Deann Christianson
Elaine M. Werner
University of the Pacific
3601 Pacific Avenue
Stockton, CA 95211

TO THE STUDENT

The purpose of this *Student Solutions Manual* is to assist you in successfully completing your mathematics course. Your class lectures are your best source of instruction and you should attend regularly. The key to success in mathematics is to regularly do the assignments given to you by your instructor.

This *Student Solutions Manual* can help you with the assignments. After you have tried and/or completed the assigned problems, verify the answers to odd problems using the answer key in the back of your textbook. You may also check your solutions against the selected exercises which have been solved for you in this manual. We have attempted to make these solutions as complete and as instructive as possible. Our solutions provide models for you in writing mathematics clearly and carefully.

We hope this manual will help you enjoy and successfully complete your mathematics course.

Contents

1 Logic

1. Statement 1: Form is "All A are B."

   Statement 2: Ansel Adams is a specific master photographer which is placed within the "master photographers" circle.

   Conclusion: VALID (x is within "artists" circle).

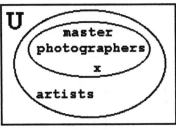

x = Ansel Adams

5. Statement 1: Form is "All A are B."

   Statement 2: Fertilizer is outside "pesticide" circle, but not necessarily outside the "environment" circle.

   Conclusion: INVALID (x could be inside the "environment" circle).

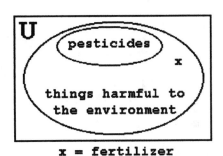

x = fertilizer

9. Statement 1: Form is "All A are B."

   Statement 2: Form is "All B are C."

   Conclusion: VALID ("all poets" are within "taxi drivers" circle).

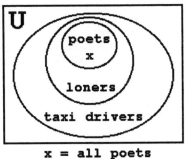

x = all poets

1

Exercise 1.1

13. Statement 1: Form is "All A are B."

    Statement 2: Route 66 is a specific road.

    Conclusion: VALID (x is within "Rome" circle).

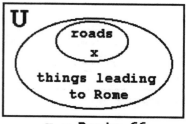

x = Route 66

17. The difference between the consecutive numbers is 6, therefore the next most likely number would probably be 21 + 6 = 27, making the sequence

    3, 3 + 6, 9 + 6, 15 + 6, _21 + 6_ = 3, 9, 15, 21, _27_ .

21. The numbers given are perfect squares, so the next most likely number would be 5^2, making the sequence

    1^2, 2^2, 3^2, 4^2, _5^2_ = 1, 4, 9, 16, _25_ .

25. The sequence looks like 3 should be added to each number but the 1 in the last position does not follow that pattern. If, however, the numbers represent the months of a year, numbered from 1 to 12, then the next number could be 4. The numbers would cycle through the months of the year representing April (4), July (7), October (10) and January (1) making the sequence

    4, 7, 10, 1, _4_ .

29. If the letters given represent the first letter of the days of the week starting with Thursday, then the next two letters would be M and T making the sequence

    Thursday, Friday, Saturday, Sunday, _M_onday, _T_uesday

33. Notice that the letter given is the letter of the alphabet that follows the first vowel that is in the word given, the letter that would fit the pattern would be "b".

37. The difference between the consecutive numbers is 3, therefore one possible number would be 11 + 3 = 14, making the sequence

    2, 2 + 3, 5 + 3, 8 + 3, _11 + 3_ = 2, 5, 8, 11, _14_ .

    Another sequence the numbers could represent would be the months of a year, then the next number would be 2 and the numbers

37. Continued

would cycle through the months of the year representing February, May, August and November.

41. Omitted.

45. Omitted.

Exercise 1.2

1. a) The given sentence "His name is George Washington" can be either true or false; therefore it **is** a statement.
   b) The given sentence is false; therefore it **is** a statement. (George Washington was the first president.)
   c) The given sentence is a question. As such, it is neither true nor false. It **is not** a statement.
   d) The given sentence is an opinion. As such, it is neither true nor false. It **is not** a statement.

5. a) $p \land q$: Statement is a conjunction (and).
   b) $\sim p \rightarrow \sim q$: Statement is a conditional (if...then...) of two negations (not).
   c) $\sim(p \lor q)$: Statement is a negation (not) of a disjunction (or).
   d) $p \land \sim q$: Statement is a conjunction (and) of a true statement and a negation (not).

9. Possible translations are:

   Translation 1: p: I do not sleep.
                    q: I drink coffee.
                    $q \rightarrow p$: (If q, then p)

   Translation 2: p: I do sleep.
                    q: I drink coffee.
                    $q \rightarrow \sim p$: (If q, then not p)

   Translation 3: p: I drink coffee.
                    q: I do sleep.
                    $p \rightarrow \sim q$: (If p, then not q)

   Translation 4: p: I drink coffee.
                    q: I do not sleep.
                    $p \rightarrow q$: (If p, then q)

Exercise 1.2

13.   Translation 1:   p: Something is an American.
                       q: It loves baseball.
                       r: It loves mom.
                       s: It loves apple pie.
                       p → (q ∧ r ∧ s): (If p, then q and r and s)
      Translation 2:   p: Something is an American.
                       q: It does not love baseball.
                       r: It loves mom.
                       s: It loves apple pie.
                       p → (~q ∧ r ∧ s): (If p, then not q and r and s)

17.   a)   p ∧ q: I am an environmentalist **and** I recycle my aluminum
           cans.
      b)   p → q: **If** I am an environmentalist, **then** I will recycle my
           aluminum cans.
      c)   ~q → ~p: **If** I do **not** recycle my aluminum cans, **then** I am
           not an environmentalist.
      d)   q ∨ ~p: I recycle my aluminum cans **or** I am **not** an
           environmentalist

21.   Omitted.

25.   Omitted.

Exercise 1.3

1.   There are n = 2 letters, so truth table rows = $2^n = 2^2 = 4$.

Label	Truth Value Assignment
p	First half (2 rows) T, last half (2 rows) F
q	Alternate T and F
~q	Opposite of q
p ∨ ~q	If p and ~q are F, then F (row 3), else T

1. Continued

	p	q	~q	p V ~q
1	T	T	F	T
2	T	F	T	T
3	F	T	F	F
4	F	F	T	T

5. There are n = 2 letters, so truth table rows = $2^n = 2^2 = 4$.

Label	Truth Value Assignment
p	First half (2 rows) T, last half (2 rows) F
q	Alternate T and F
~q	Opposite of q
p → ~q	If p is T and ~q is F, then F (row 1), else T

	p	q	~q	p → ~q
1	T	T	F	F
2	T	F	T	T
3	F	T	F	T
4	F	F	T	T

Exercise 1.3

9.   There are n = 2 letters, so truth table rows = 2^n = 2^2 = 4.

Label	Truth Value Assignment
p	First half (2 rows) T, last half (2 rows) F
q	Alternate T and F
~p	Opposite of p
p V q	If p and q are F, then F (row 4), else T
(p V q) → ~p	If (p V q) is T and ~p is F, then F (rows 1,2), otherwise T

	p	q	~p	p V q	(p V q) → ~p
1	T	T	F	T	F
2	T	F	F	T	F
3	F	T	T	T	T
4	F	F	T	F	T

13.   There are n = 3 letters, so truth table rows = 2^n = 2^3 = 8.

Label	Truth Value Assignment
p	First half (4 rows) T, last half (4 rows) F
q	Alternate 2 Ts and 2 Fs
r	Alternate T and F
q V r	If q and r are F, then F (rows 4,8), else T
~(q V r)	Opposite of (q V r)
p ∧ ~(q V r)	If p and ~(q V r) are T, then T (row 4), otherwise F

13. Continued

	p	q	r	q $\vee$ r	~(q $\vee$ r)	p $\wedge$ ~(q $\vee$ r)
1	T	T	T	T	F	F
2	T	T	F	T	F	F
3	T	F	T	T	F	F
4	T	F	F	F	T	T
5	F	T	T	T	F	F
6	F	T	F	T	F	F
7	F	F	T	T	F	F
8	F	F	F	F	T	F

17. There are n = 3 letters, so truth table rows = $2^n = 2^3 = 8$.

Label	Truth Value Assignment
p	First half (4 rows) T, last half (4 rows) F
q	Alternate 2 Ts and 2 Fs
r	Alternate T and F
~r	Opposite of r
~r $\vee$ p	If ~r and p are F, then F (rows 5,7), otherwise T
q $\wedge$ p	If q and p are T, then T (rows 1,2), else F
(~r $\vee$ p) $\rightarrow$ (q $\wedge$ p)	If (~r $\vee$ p) is T and (q $\wedge$ p) is F, then F (rows 3,4,6,8), else T

7

Exercise 1.3

17. Continued

	p	q	r	~r	~r V p	q ∧ p	(~r V p) → (q ∧ p)
1	T	T	T	F	T	T	T
2	T	T	F	T	T	T	T
3	T	F	T	F	T	F	F
4	T	F	F	T	T	F	F
5	F	T	T	F	F	F	T
6	F	T	F	T	T	F	F
7	F	F	T	F	F	F	T
8	F	F	F	T	T	F	F

21. p: It is raining.
    q: The streets are wet.
    p → q: If it is raining, then the streets are wet.

There are n = 2 letters, so truth table rows = $2^n = 2^2 = 4$.

Label	Truth Value Assignment
p	First half (2 rows) T, last half (2 rows) F
q	Alternate T and F
p → q	If p is T and q is F, then F (row 2), otherwise T

	p	q	p → q
1	T	T	T
2	T	F	F
3	F	T	T
4	F	F	T

25. p: Leaded gasoline is used.
     q: The catalytic converter is damaged.
     r: The air is polluted.
     $p \to (q \wedge r)$: If leaded gasoline is used, then the catalytic
                   converter is damaged and the air is polluted.

There are n = 3 letters, so truth table rows = $2^n = 2^3 = 8$.

Label	Truth Value Assignment
p	First half (4 rows) T, last half (4 rows) F
q	Alternate 2 Ts and 2 Fs
r	Alternate T and F
$q \wedge r$	If q and r are T, then T (rows 1,5), otherwise F
$p \to (q \wedge r)$	If p is T and $(q \wedge r)$ is F, then F (rows 2,3,4), otherwise T

	p	q	r	$q \wedge r$	$p \to (q \wedge r)$
1	T	T	T	T	T
2	T	T	F	F	F
3	T	F	T	F	F
4	T	F	F	F	F
5	F	T	T	T	T
6	F	T	F	F	T
7	F	F	T	F	T
8	F	F	F	F	T

29. p: The streets are wet
     q: It is raining.
     p ∨ ~q: The streets are wet or it is not raining.
     q → p: If it is raining, then the streets are wet.

     There are n = 2 letters, so truth table rows = $2^n = 2^2 = 4$.

Label	Truth Value Assignment
p	First half (2 rows) T, last half (2 rows) F
q	Alternate T and F
~q	Opposite of q
p ∨ ~q	If p and ~q are F, then F (row 3), else T
q → p	If q is T and p is F, then F (row 3), otherwise T

	p	q	~q	p ∨ ~q	q → p
1	T	T	F	T	T
2	T	F	T	T	T
3	F	T	F	F	F
4	F	F	T	T	T

p ∨ ~q ≡ q → p. The two statements are **equivalent** because the values in the columns labeled (p ∨ ~q) and (q → p) are the same.

33. p: Handguns are outlawed.
     q: Outlaws have handguns.
     p → q: If handguns are outlawed, then outlaws have handguns.
     q → p: If outlaws have handguns, then handguns are outlawed.

     There are n = 2 letters, so truth table rows = $2^n = 2^2 = 4$.

Label	Truth Value Assignment
p	First half (2 rows) T, last half (2 rows) F
q	Alternate T and F
p → q	If p is T and q is F, then F (row 2), otherwise T
q → p	If q is T and p is F, then F (row 3), otherwise T

33. Continued

	p	q	p → q	q → p
1	T	T	T	T
2	T	F	F	T
3	F	T	T	F
4	F	F	T	T

The two statements are **not equivalent** because the values in the columns labeled (p → q) and (q → p) are **not** the same.

37. p: The plaintiff is innocent.
    q: The insurance company settles out of court.
    p ∨ ~q: The plaintiff is innocent or the insurance company does not settle out of court.
    q ∧ ~p: The insurance company settles out of court and the plaintiff is not innocent.

There are n = 2 letters, so truth table rows = $2^n = 2^2 = 4$.

Label	Truth Value Assignment
p	First half (2 rows) T, last half (2 rows) F
q	Alternate T and F
~p	Opposite of p
~q	Opposite of q
p ∨ ~q	If p and ~q are F, then F (row 3), else T
q ∧ ~p	If q and ~p are T, then T (row 3), else F

	p	q	~p	~q	p ∨ ~q	q ∧ ~p
1	T	T	F	F	T	F
2	T	F	F	T	T	F
3	F	T	T	F	F	T
4	F	F	T	T	T	F

37. Continued

    The two statements are **not** equivalent because the values in the columns labeled (p ∨ ~q) and (q ∧ ~p) are **not** the same.

41. p: I have a college degree.
    q: I am not employed
    p ∧ q: I have a college degree and I am not employed.

    The negation is: ~(p ∧ q) ≡ ~p ∨ ~q
        ~p: I do not have a college degree.
        ~q: I am employed.
        ~p ∨ ~q: I do not have a college degree or I am employed.

45. p: The building contains asbestos.
    q: The original contractor is responsible.
    p → q: If the building contains asbestos, then the original
        contractor is responsible.

    The negation is: ~(p → q) ≡ p ∧ ~q
        p ∧ ~q: The building contains asbestos and the original
                contractor is not responsible.

49. Omitted.

53. Omitted.

Exercise 1.4

1. a) p → q: If she is a police officer, then she carries a gun.
    b) q → p: If she carries a gun, then she is a police officer.
    c) ~p → ~q: If she is not a police officer, then she does not
            carry a gun.
    d) ~q → ~p: If she does not carry a gun, then she is not a
            police officer.
    e) Part a is equivalent to part d and part b is equivalent to
        part c because each equivalent pair was formed by negating
        and interchanging the premise and the conclusion.

5. p: You will pass this mathematics course.
    q: You will fulfill a graduation requirement.

    a) The inverse is ~p → ~q: If you do **not** pass this mathematics
        course, then you will **not** fulfill a graduation requirement.
    b) The converse is q → p: If you fulfill a graduation
        requirement, then you will pass this mathematics course.
    c) The contrapositive is ~q → ~p: If you do **not** fulfill a
        graduation requirement, then you will **not** pass this
        mathematics course.

9. p: You do not eat meat.
    q: You are a vegetarian.

    a) The inverse is ~p → ~q: If you **do** eat meat, then you are
        not a vegetarian.
    b) The converse is q → p: If you are a vegetarian, then you do
        not eat meat.
    c) The contrapositive is ~q → ~p: If you are **not** a vegetarian,
        then you **do** eat meat.

13. a) In an "only if" compound statement the conclusion follows
        the "only if" so the premise is "I buy foreign products" and
        the conclusion is "domestic products are not available".
    b) If I buy foreign products, then domestic products are not
        available.
    c) A truth table needs to be constructed to find the conditions
        that make the statement false.

        p: I buy foreign products.
        q: Domestic products are available.
        p → ~q: If I buy foreign products, then domestic
            products are not available.

    There are 2 letters, so truth table rows = $2^n = 2^2 = 4$.

Label	Truth Value Assignment
p	First half (2 rows) T, last half (2 rows) F
q	Alternate T and F
~q	Opposite of q
p → ~q	If p is T and ~q is F, then F (row 1), else T

Exercise 1.4

13.c) Continued

	p	q	~q	p → ~q
1	T	T	F	F
2	T	F	T	T
3	F	T	F	T
4	F	F	T	T

The conditions given in row 1 (p and q both true) would give a false expression. Therefore, the statement "I buy foreign products only if domestic products are not available," is false when I buy foreign products and domestic products are available.

17. p: You obtain a refund.
q: You have a receipt.
p → q: If you obtain a refund, then you have a receipt.
q → p: If you have a receipt, then you will obtain a refund.

The biconditional is p ↔ q ≡ [(p → q) ∧ (q → p)]: If you obtain a refund, then you have a receipt, **and** if you have a receipt, then you will obtain a refund.

21. p: A polygon is a triangle.
q: The polygon has three sides.
p → q: If a polygon is a triangle, then the polygon has three sides.
q → p: If the polygon has three sides, then the polygon is a triangle.

The biconditional is p ↔ q ≡ [(p → q) ∧ (q → p)]: If a polygon is a triangle, then the polygon has three sides, **and** if the polygon has three sides then the polygon is a triangle.

25. p: You earn less than $12,000 per year.
q: You are eligible for assistance.
p → q: If you earn less than $12,000 per year, then you are eligible for assistance.
~q → ~p: If you are **not** eligible for assistance, then you earn **at least** $12,000 per year.

25. **Continued**

There are $n = 2$ letters, so truth table rows $= 2^n = 2^2 = 4$.

Label	Truth Value Assignment
p	First half (2 rows) T, last half (2 rows) F
q	Alternate T and F
p → q	If p is T and q is F, then F (row 2), otherwise T
~q	Opposite of q
~p	Opposite of p
~q → ~p	If ~q is T and ~p is F, then F (row 2), otherwise T

	p	q	p → q	~q	~p	~q → ~p
1	T	T	T	F	F	T
2	T	F	F	T	F	F
3	F	T	T	F	T	T
4	F	F	T	T	T	T

p → q ≡ ~q → ~p. The two statements are **equivalent** because the values in columns (p → q) and (~q → ~p) are the same.

29. p: It is not raining.
    q: I walk to work.
    p → q: If it is not raining, then I walk to work.

   (p → q) ≡ (~q → ~p)   (A conditional is equivalent to its
      contrapositive, therefore we need to negate and interchange
      the premise and the conclusion.)

   ~q → ~p: If I do **not** walk to work, then it **is** raining.

Exercise 1.4

33. p: You eat meat.
    q: You are not a vegetarian.
    p → q: If you eat meat, then you are not a vegetarian.

    (p → q) ≡ (~q → ~p)   (A conditional is equivalent to its
        contrapositive, therefore we need to negate and interchange
        the premise and conclusion.)

    ~q → ~p: If you **are** a vegetarian, then you do **not** eat meat.

37. p: It is Sunday.
    q: I go to church.

    i)   p → q: If it is Sunday, then I go to church.
    ii)   q → p: I go to church **only if** it is Sunday. (Conclusion
          follows "only if".)
    iii) ~q → ~p: If I do **not** go to church, then it is **not** Sunday.
    iv)   ~p → ~q: If it is **not** Sunday, then I do **not** go to church.

    Using Figure 1.40:
            i ≡ iii   conditional ≡ contrapositive
            ii ≡ iv   inverse ≡ converse

41. Omitted.

Exercise 1.5

1.   1. p → q   }   hypothesis
     2. p   }
     ∴ q   }   conclusion

     [(p → q) ∧ p] → q }   conditional representation

5.   1. p → q   }   hypothesis
     2. ~p   }
     ∴ ~q   }   conclusion

     [(p → q) ∧ ~p] → ~q }   conditional representation

9. 1. p → q   }   hypothesis (Hyp)
    <u>2. ~q</u>
    ∴ ~p   }   conclusion (Concl)

    [(p → q) ∧ ~q] → ~p }   conditional representation (Cond)

There are n = 2 letters, so truth table rows = $2^n = 2^2 = 4$.

Label	Truth Value Assignment
p	First half (2 rows) T, last half (2 rows) F
q	Alternate T and F
Hyp #1: p → q	If p is T and q is F, then F (row 2), otherwise T
Hyp #2: ~q	Opposite of q
All: #1 ∧ #2 (p → q) ∧ ~q	If (p → q) and ~q are T, then T (row 4), otherwise F
Conclusion: ~p	Opposite of p
Conditional: [(p → q) ∧ ~q] → ~p	If (1 ∧ 2) is T and the Conclusion is F, then F (no rows), otherwise T

			#1	#2	All	Concl	Cond
	p	q	p → q	~q	1 ∧ 2	~p	[1 ∧ 2] → ~p
1	T	T	T	F	F	F	T
2	T	F	F	T	F	F	T
3	F	T	T	F	F	T	T
4	F	F	T	T	T	T	T

The argument is **valid** as the conditional [(p → q) ∧ ~q] → ~p is always true.

17

Exercise 1.5

13. p: The Democrats have a majority.
     q: Smith is appointed.
     r: Student loans are funded.

Symbolic form of argument is:

$$1.\ p \to (q \wedge r)$$
$$2.\ q \vee \sim r$$ hypothesis
$$\therefore\ \sim p$$ conclusion

$\{[p \to (q \wedge r)] \wedge (q \vee \sim r)\} \to \sim p \}$ conditional representation

There are n = 3 letters, so truth table rows = $2^n = 2^3 = 8$.

Label	Truth Value Assignment
p	First half (4 rows) T, last half (4 rows) F
q	Alternate 2 Ts and 2 Fs
r	Alternate T and F
~r	Opposite of r
q ∧ r	If q and r are T, then T (rows 1,5), else F
Hyp #1: p → (q ∧ r)	If p is T and (q ∧ r) is F, then F (rows 2,3,4), otherwise T
Hyp #2: q ∨ ~r	If q is F and ~r is F then F (rows 3,7), otherwise T
All: 1 ∧ 2	If #1 is T and #2 is T, then T (rows 1,5,6,8), otherwise F
Conclusion: ~p	Opposite of p
Conditional: (See Above)	If All is T and ~p is F, then F (row 1), otherwise T

13.   Continued

	p	q	r	~r	q ∧ r	#1	#2	All	Concl	Cond
1	T	T	T	F	T	T	T	T	F	F
2	T	T	F	T	F	F	T	F	F	T
3	T	F	T	F	F	F	F	F	F	T
4	T	F	F	T	F	F	T	F	F	T
5	F	T	T	F	T	T	T	T	T	T
6	F	T	F	T	F	T	T	T	T	T
7	F	F	T	F	F	T	F	F	T	T
8	F	F	F	T	F	T	T	T	T	T

The argument is **invalid** because row 1 of the conditional column of the truth table is false. If all three premises (p, q and r) are true, then the Democrats must have a majority.

17.   p: Someone is a lawyer.
q: You study logic.
r: You are a scholar.

Symbolic form of argument is:

1. p → q
2. q → r   } hypothesis
3. ~r
∴   ~p   } conclusion

[(p → q) ∧ (q → r) ∧ (~r)] → ~p) }conditional representation

17. Continued

There are n = 3 letters, so truth table rows = $2^n = 2^3 = 8$.

Label	Truth Value Assignment
p	First half (4 rows) T, last half (4 rows) F
q	Alternate 2 Ts and 2 Fs
r	Alternate T and F
Hyp #1: p → q	If p is T and q is F, then F (rows 3,4), otherwise T
Hyp #2: q → r	If q is T and r is F, then F (rows 2,6), otherwise T
Hyp #3: ~r	Opposite of r
All: 1 ∧ 2 ∧ 3	If #1, #2, and #3 are **all** T, then T (row 8), otherwise F
Conclusion: ~p	Opposite of p
Conditional: (See Above)	If All is T and Conclusion is F, then F (no rows), otherwise T

	p	q	r	#1	#2	#3	All	Concl	Cond
1	T	T	T	T	T	F	F	F	T
2	T	T	F	T	F	T	F	F	T
3	T	F	T	F	T	F	F	F	T
4	T	F	F	F	T	T	F	F	T
5	F	T	T	T	T	F	F	T	T
6	F	T	F	T	F	T	F	T	T
7	F	F	T	T	T	F	F	T	T
8	F	F	F	T	T	T	T	T	T

The argument is **valid** because the conditional column of the truth table is always true.

21. 1. If you are not in a hurry, you eat at Lulu's Diner.
     2. If you are in a hurry, you do not eat good food.
     <u>3. You eat at Lulu's.</u>
     Therefore, you eat good food.

     p: You are in a hurry.
     q: You eat at Lulu's Diner.
     r: You eat good food.

     Symbolic form of argument is:

     1. ~p → q ⎫
     2. p → ~r ⎬   hypothesis
     <u>3. q</u>
     ∴ r   ⎬   conclusion

     [(~p → q) ∧ (p → ~r) ∧ q] → r   ⎬   conditional representation

     There are n = 3 letters, so truth table rows = $2^n = 2^3 = 8$.

Label	Truth Value Assignment
p	First half (4 rows) T, last half (4 rows) F
Hyp #3: q	Alternate 2 Ts and 2 Fs
r	Alternate T and F
~p	Opposite of p
~r	Opposite of r
Hyp #1: ~p → q	If ~p is T and q is F, then F (rows 7,8), otherwise T
Hyp #2: p → ~r	If p is T and ~r is F, then F (rows 1,3), otherwise T
All: 1 ∧ 2 ∧ 3	If #1, #2 and #3 are **all** T, then T (rows 2,5,6), otherwise F
Conclusion: r	Alternate T and F
Conditional: (See Above)	If (1 ∧ 2 ∧ 3) is T and r is F, then F (rows 2,6), otherwise T

21. Continued

		#3				#1	#2	All	Conc	Cond
	p	q	r	~p	~r	~p → q	p → ~r	1 ∧ 2 ∧ 3	r	Cond
1	T	T	T	F	F	T	F	F	T	T
2	T	T	F	F	T	T	T	T	F	F
3	T	F	T	F	F	T	F	F	T	T
4	T	F	F	F	T	T	T	F	F	T
5	F	T	T	T	F	T	T	T	T	T
6	F	T	F	T	T	T	T	T	F	F
7	F	F	T	T	F	F	T	F	T	T
8	F	F	F	T	T	F	T	F	F	T

The argument is **invalid** because the conditional statements in rows 2 and 6 of the truth table are false. In both cases you are eating at Lulu's Diner, but it could be false that you are eating good food or that you are in a hurry.

25. Omitted.

29. Omitted.

Chapter 1 Review

1. Omitted.

5. Statement 1: Form is "All A are B."

   Statement 2: Casey Jones is outside the "engineers" circle.

   Conclusion: VALID (*x* cannot be inside "mechanics" circle).

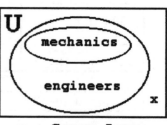

x = Casey Jones

9. a) p ∧ q: Statement is a conjunction (and).

    b) ~q → ~p: Statement is a conditional (if...then...) of two
        negations (not).

    c) p ∧ ~q): Statement is a conjunction (and) of a true
        statement and a negation (not).

    d) ~(p ∨ q): Statement is a negation (not) of a disjunction
        (or).

13. There are n = 2 letters, so truth table rows = $2^n = 2^2 = 4$.

Label	Truth Value Assignment
p	First half (2 rows) T, last half (2 rows) F
q	Alternate T and F
~q	Opposite of q
p ∨ ~q	If p and ~q are F, then F (row 3), otherwise T

	p	q	~q	p ∨ ~q
1	T	T	F	T
2	T	F	T	T
3	F	T	F	F
4	F	F	T	T

17. There are n = 3 letters, so truth table rows = $2^n = 2^3 = 8$.

Label	Truth Value Assignment
p	First half (4 rows) T, last half (4 rows) F
q	Alternate 2 Ts and 2 Fs
r	Alternate T and F
p ∨ r	If p and r are both F, then F (rows 6,8), otherwise T
~(p ∨ r)	Opposite of (p ∨ r)
q ∨ ~(p ∨ r)	If q and ~(p ∨ r) are both F, then F (rows 3,4,7), otherwise T

17. Continued

	p	q	r	p $\vee$ r	~(p $\vee$ r)	q $\vee$ ~(p $\vee$ r)
1	T	T	T	T	F	T
2	T	T	F	T	F	T
3	T	F	T	T	F	F
4	T	F	F	T	F	F
5	F	T	T	T	F	T
6	F	T	F	F	T	T
7	F	F	T	T	F	F
8	F	F	F	F	T	T

21. p: The car is reliable.
     q: The car is expensive.
     ~p $\vee$ q: The car is unreliable or expensive.
     p $\rightarrow$ q: If the car is reliable, then it is expensive.

     There are n = 2 letters, so truth table rows = $2^n = 2^2 = 4$.

Label	Truth Value Assignment
p	First half (2 rows) T, last half (2 rows) F
q	Alternate T and F
~p	Opposite of p
~p $\vee$ q	If ~p and q are both F, then F (row 2), otherwise T
p $\rightarrow$ q	If p is T and q if F, then F (row 2), otherwise T

21. Continued

	p	q	~p	~p ∨ q	p → q
1	T	T	F	T	T
2	T	F	F	F	F
3	F	T	T	T	T
4	F	F	T	T	T

~p ∨ q ≡ p → q. The two statements are **equivalent** because the values in the columns labeled (~p ∨ q) and (p → q) are the same.

25. p: Jesse had a party.
    q: Nobody came to the party.
    p ∧ q: Jesse had a party and nobody came.

    By De Morgan's Law the negation is: ~(p ∧ q) ≡ ~p ∨ ~q
        ~p: Jesse did not have a party.
        ~q: Somebody came to the party.
        ~p ∨ ~q: Jesse did not have a party or someone came to the party.

29. p: His application is ignored.
    q: The selection procedure has been violated.
    p → q: If his application is ignored, the selection procedure has been violated.

    The negation is: ~(p → q) ≡ p ∧ ~q
        ~q: The selection procedure has **not** been violated.
        p ∧ ~q: His application is ignored and the selection procedure has not been violated.

33. a) In an "only if" compound statement the conclusion follows the "only if" so the premise is "The economy improves" and the conclusion is "unemployment goes down"
    b) If the economy improves, then unemployment goes down.

37. p: It is **not** raining.
     q: I ride my bicycle to work.

     i)   p → q: If it is not raining, I ride my bicycle to work.
     ii) q → p: If I ride my bicycle to work, it is not raining.
     iii) ~q → ~p: If I do **not** ride my bicycle to work, it **is** raining.
     iv) ~p → ~q: If it **is** raining, I do **not** ride my bicycle to work.

     Using Figure 1.40:   i ≡ iii   conditional ≡ contrapositive
                          ii ≡ iv   inverse ≡ converse

41. p: The Republicans have a majority.
     q: Farnsworth is appointed.
     r: New taxes are imposed.

     Symbolic form of argument is:

          1. p → (q ∧ ~r)   }   hypothesis
          2. r
          ∴ ~p ∨ ~q   }   conclusion

          { [p → (q ∧ ~r)] ∧ r} → (~p ∨ ~q) } conditional representation

     There are n = 3 letters, so truth table rows = $2^n = 2^3 = 8$.

Label	Truth Value Assignment
p	First half (4 rows) T, last half (4 rows) F
q	Alternate 2 Ts and 2 Fs
Hyp #2: r	Alternate T and F
~p	Opposite of p
~q	Opposite of q
~r	Opposite of r
q ∧ ~r	If q is T and ~r is T, then T (rows 2,6), otherwise F
Hyp #1: p → q ∧ ~r	If p is T and q ∧ ~r is F, then F (rows 1,3,4), otherwise T
1 ∧ 2	If #1 is T and #2 is T, then T (rows 5,7), otherwise F

41. Continued

Conclusion: ~p ∨ ~q	If ~p is F and ~q is F, then F (rows 1,2), otherwise T
Conditional: (See Above)	If (1 ∧ 2) is T and (~p ∨ ~q) is F, then F (no rows), otherwise T

			#2					#1	All		
	p	q	r	~p	~q	~r	q ∧ ~r	#1	1 ∧ 2	Conc	Cond
1	T	T	T	F	F	F	F	F	F	F	T
2	T	T	F	F	F	T	T	T	F	F	T
3	T	F	T	F	T	F	F	F	F	T	T
4	T	F	F	F	T	T	F	F	F	T	T
5	F	T	T	T	F	F	F	T	T	T	T
6	F	T	F	T	F	T	T	T	F	T	T
7	F	F	T	T	T	F	F	T	T	T	T
8	F	F	F	T	T	T	F	T	F	T	T

The argument is **valid** as the conditional
{[p → (q ∧ ~r)] ∧ r} → (~p ∨ ~q) is always true.

45. 1. If our oil supply is cut off, our economy collapses.
    2. If we go to war, our economy does not collapse.
    Therefore, if our oil supply is not cut off, we do not go to war.

    p: Our oil supply is cut off.
    q: Our economy collapses.
    r: We go to war.

    Symbolic form of argument is:

    1. p → q   } hypothesis
    2. r → ~q
    ∴ ~p → ~r   } conclusion

    [(p → q) ∧ (r → ~q)] → (~p → ~r)   } conditional
                                            representation

45. Continued
    There are n = 3 letters, so truth table rows = $2^n = 2^3 = 8$.

Label	Truth Value Assignment
p	First half (4 rows) T, last half (4 rows) F
q	Alternate 2 Ts and 2 Fs
r	Alternate T and F
~p	Opposite of p
~q	Opposite of q
~r	Opposite of r
Hyp #1: p → q	If p is T and q is F, then F (rows 3,4), otherwise T
Hyp #2: r → ~q	If r is T and ~q is F, then F (rows 1,5), otherwise T
1 ∧ 2	If #1 and #2 are T, then T (rows 2,6,7,8), otherwise F
Conclusion: ~p → ~r	If ~p is T and ~r is F, then F (rows 5,7), otherwise T
Conditional: (See Above)	If (1 ∧ 2) is T and (~p → ~r) is F, then F (row 7), otherwise T

	p	q	r	~p	~q	~r	#1	#2	All	Concl	Cond
1	T	T	T	F	F	F	T	F	F	T	T
2	T	T	F	F	F	T	T	T	T	T	T
3	T	F	T	F	T	F	F	T	F	T	T
4	T	F	F	F	T	T	F	T	F	T	T
5	F	T	T	T	F	F	T	F	F	F	T
6	F	T	F	T	F	T	T	T	T	T	T
7	F	F	T	T	T	F	T	T	T	F	F
8	F	F	F	T	T	T	T	T	T	T	T

The argument is **invalid** because row 7 of the conditional column in the truth table is false.

2   Sets and Counting

Exercise 2.1

1. a) "The set of all black automobiles" **is** well-defined because
      an automobile is either black or not black.
   b) "The set of all inexpensive automobiles" **is not** well-defined
      because we need a level of cost (e.g., under $8,000)
      specified to determine whether an automobile is expensive or
      inexpensive.
   c) "The set of all prime numbers" **is** well-defined as it can be
      determined whether or not a number is prime or not prime.
   d) "The set of all large numbers" **is not** well-defined because
      different people will have different definitions for large.

5. Proper subsets:

   > No elements: ∅
   > One element: {yes}, {no}, {undecided}
   > Two elements: {yes, no}, {yes, undecided}, {no, undecided}

   Improper subsets:
   > Three elements: {yes, no, undecided}

9. B′ = set of all elements not in B
      = {Monday, Tuesday, Wednesday, Thursday}

13. A ∩ B (A intersect B) is the region that is common to both A
    and B.

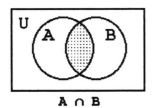

A ∩ B

17. A ∪ B' (A union the complement of B) is the region that is in A together with all the region that is not in B.

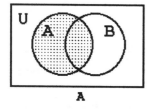

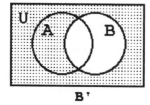

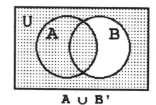

21. A' ∪ B' (the complement of A union the complement of B) is the region that is not in A together with the region that is not in B.

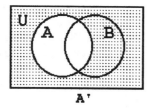

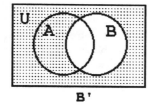

25. U = {students|the student is a high school senior}
    A = {students|the student owns an automobile}
    M = {students|the student owns a motorcycle}
    A ∩ M = {students|the student owns both}

    n(U) = 500, n(A) = 91, n(M) = 123, n(A ∩ M) = 29

    a)   n(A only) = 91 - 29 = 62
         n(M only) = 123 - 29 = 94
         n(A ∪ M) = n(A) + n(M) - n(A ∩ M)
                  = 91 + 123 - 29 = 185
         n(None) = n(U) - n(A ∪ M) = 500 - 185 = 315

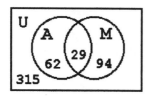

25. Continued

    b)   The percent of students owning an automobile or a motorcyle
         is

    $$\frac{n(A \cup M)}{n(U)} = \frac{185}{500} = 0.37 = 37\%$$

29. M = {Maine, Maryland, Massachusetts, Michigan, Minnesota,
    Mississippi, Missouri, Montana}   (Hint: Use an atlas.)

    n(U) = 50, n(M) = 8

    n(M′) = n(U) − n(M) = 50 − 8 = 42

33. R = {September, October, November, December}.

    n(U) = 12, n(R) = 4

    n(R′) = n(U) − n(R) = 12 − 4 = 8

37. U = {cards|a card is from an ordinary deck}
    S = {cards|a card is a spade}
    A = {cards|a card is an ace}
    S ∩ A = {cards|a card is both a spade and an ace}

    n(U) = 52, n(S) = 13, n(A) = 4, n(S ∩ A) = 1

    Find "spades **or** aces" which is the union of set S and set A.

    $$\begin{aligned}
    n(S \cup A) &= n(S) + n(A) - n(S \cap A) \\
    &= 13 + 4 - 1 \\
    &= 16 \text{ spades or aces}
    \end{aligned}$$

41. U = {cards|a card is from an ordinary deck}
    F = {cards|a card is a face card (jack, queen, king)}
    B = {cards|a card is black (spade, club)}

    Find "face cards **and** black" which is the intersection of set F
    and set B.

    F ∩ B = {club jack, club queen, club king, spade jack,
             spade queen, spade king}

    n(F ∩ B) = 6 face cards that are black

Exercise 2.1

45. U = {cards|a card is from an ordinary deck}
    A = {cards|a card is an ace}
    E = {cards|a card is an eight}

    Find "aces **and** eights" which is the intersection of set A and set E.

    $A \cap E = \emptyset$ There are no cards that are both an ace and an eight.

    $n(A \cap E) = 0$

49. a)   A = {a}, n(A) = 1.
         S = the subsets of A = {∅, {a}}
         n(S) = 2

    b)   A = {a,b}, n(A) = 2.
         S = the subsets of A = {∅, {a}, {b}, {a,b}}
         n(S) = 4

    c)   A = {a,b,c}, n(A) = 3
         S = the subsets of A = {∅, {a}, {b}, {c}, {a,b}, {a,c},
             {b,c}, {a,b,c,}}
         n(S) = 8

    d)   A = {a,b,c,d}, n(A) = 4.
         S = the subsets of A = {∅, {a}, {b}, {c}, {d}, {a,b},
             {a,c}, {a,d}, {b,c}, {b,d}, {c,d}, {a,b,c,},
             {a,b,d}, {a,c,d}, {b,c,d}, {a,b,c,d}}
         n(S) = 16

    e)   Yes, each time an element is added to the set the number of subsets doubles. In other words, the number of subsets is two raised to the cardinal number $(2^{n(A)})$.

    f)   A = {a,b,c,d,e,f}, n(A) = 6
         The number of subsets would be $2^{n(A)} = 2^6 = 64$.

53. Omitted.

57. Omitted.

61. Comparing the possible answers against the given restrictions leaves (d) as the only answer that conforms to all the restrictions.

Answer	John	Juneko	Restriction Violated
a.	J, K, L	M, N, O	M and O cannot be in the same group.
b.	J, K, P	L, M, N	J and P cannot be in the same group.
c.	K, N, P	J, M, O	If N is in John's group, P must be in Juneko's group.
d.	L, M, N	K, O, P	No violation of the restrictions
e.	M, O, P	J, K, N	M and O cannot be in the same group

65. The answer must be (e). By the restrictions: If K is in John's group, M must be in Juneko's group. Also, M and O cannot be in the same group, therefore O cannot be in Juneko's group. O could be in John's group, but it could also be the poster that is not used. Nothing is known about J, L, or N which means answer (e) is the only one that can be true.

Exercise 2.2

1. U = {persons|the person was interviewed in the survey}
    V = {persons|the person owned a VCR}
    M = {persons|the person owned a microwave oven}

    n(U) = 200, n(V) = 94, n(M) = 127, n(V ∩ M) = 78

Region	Set	n(Set)	Calculation
Both	V ∩ M	78	Given
V Only	V ∩ M'	16	n(V) – n(V ∩ M) = 94 – 78 = 16
M Only	M ∩ V'	49	n(M) – n(V ∩ M) = 127 – 78 = 49
Neither	(V ∪ M)'	57	n(U) – n(V ∩ M) – N(V ∩ M') – n((M ∩ V') = 200 – 78 – 16 – 49 = 57

Construct a Venn diagram from the table.

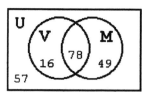

a) n(V ∪ M) = 16 + 78 + 49 = 143 people who own **either** a VCR **or** a microwave oven.
b) n(V ∩ M') = 16 people who own **only a VCR** but **not** a microwave oven.
c) n(M ∩ V') = 49 people who own **only a microwave oven** but **not** a VCR.
d) n((V ∪ M)') = 57 people who own **neither** a VCR **nor** a microwave oven.

5. U = {adults|the adult was in the survey}
    B = {adults|the adult watched the Big Game}
    M = {adults|the adult watched the New Movie}

n(U) = 674, n((B ∪ M)') = 226, n(M) = 289, n(M ∩ B') = 183

Region	Set	n(Set)	Calculation
Neither	(B ∪ M)'	226	Given
M Only	M ∩ B'	183	Given
Both	B ∩ M	106	n(M) − n(M ∩ B') = 289 − 183 = 106
B Only	B ∩ M'	159	n(U) − n(M) − n((B ∪ M)') = 674 − 289 − 226 = 159

Construct a Venn diagram from the table.

a) n(B ∩ M) = 106 adults watched **both** programs.
b) n(B ∪ M) = 159 + 106 + 183 = 448 adults watched **at least one** program.
c) n(B) = 159 + 106 = 265 adults watched the Big Game.
d) n(B ∩ M') = 159 adults watched **only the Big Game** but **not** the New Movie.

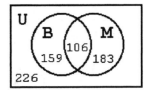

Exercise 2.2

9. U = {persons|the person was interviewed in the survey}
   M = {persons|the person owned a microwave oven}
   A = {persons|the person owned an answering machine}
   V = {persons|the person owned a VCR}

   n(M) = 313,
   n(A) = 232,
   n(V) = 269,
   n(had all three) = n(M ∩ A ∩ V) = 69,
   n(had none) = n(M' ∩ A' ∩ V') = 64,
   n(had A and V) = n(A ∩ V) = 98,
   n(had A only) = n(M' ∩ A ∩ V') = 57,
   n(had M and V but not A) = n(M ∩ A' ∩ V) = 104

Region	Set	n(Set)	Calculation
All	M ∩ A ∩ V	69	Given
None	M' ∩ A' ∩ V'	64	Given
A Only	M' ∩ A ∩ V'	57	Given
A and V and not M	M' ∩ A ∩ V	29	n(A ∩ V) − n(M ∩ A ∩ V) = 98 − 69 = 29
A and M and not V	M ∩ A ∩ V'	77	n(A) − n(M ∩ A ∩ V) − n(M' ∩ A ∩ V) − n(M' ∩ A ∩ V') = 232 − 69 − 29 − 57 = 77
M and V and not A	M ∩ A' ∩ V	104	Given
M only	M ∩ A' ∩ V'	63	n(M) − n(M ∩ A ∩ V) − n(M ∩ A ∩ V') − n(M ∩ A' ∩ V) = 313 − 69 − 77 − 104 = 63
V only	M' ∩ A' ∩ V	67	n(V) − n(M ∩ A ∩ V) − n(M' ∩ A ∩ V) − n(M ∩ A' ∩ V) = 269 − 69 − 29 − 104 = 67

9.   Continued

Construct a Venn diagram from the table.

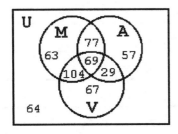

n(U) = 69 + 64 + 57 + 29 + 77 + 104 + 63 + 67 = 530

a)   n(Microwave only) = n(M ∩ A′ ∩ V′) = 63

In percent, $\dfrac{n(M \cap A' \cap V')}{n(U)} = \dfrac{63}{530} = 0.118868 \approx 12\%$

b)   n(VCR only) = n(M′ ∩ A′ ∩ V)

In percent, $\dfrac{n(M' \cap A' \cap V)}{n(U)} = \dfrac{67}{530} = 0.126415 \approx 13\%$

13.   U = {members|the member belongs to the Eye and I Photo Club}
B = {members|the member used black and white film}
C = {members|the member used color film}
I = {members|the member used infrared film}

n(B) = 77,
n(B only) = n(B ∩ C′ ∩ I′) = 24,
n(C) = 65,
n(C only) = n(C ∩ B′ ∩ I′) = 18,
n(B or C) = n(B ∪ C) = 101,
n(I) = 27,
n(all) = n(B ∩ C ∩ I) = 9,
n(none) = n(B′ ∩ C′ ∩ I′) = 8

Exercise 2.2

13.   Continued

Region	Set	n(Set)	Calculation
None	B' ∩ C' ∩ I'	8	Given
All	B ∩ C ∩ I	9	Given
B Only	B ∩ C' ∩ I'	24	Given
C Only	B' ∩ C ∩ I'	18	Given
B and C and not I	B ∩ C ∩ I'	32	Step 1: $n(B \cup C) = n(B) + n(C)$ $\qquad\qquad - n(B \cap C)$ $101 = 77 + 65 - n(B \cap C)$ $n(B \cap C) = 142 - 101 = 41$ Step 2: $n(B \cap C) - n(B \cap C \cap I)$ $= 41 - 9 = 32$
B and I and not C	B ∩ C' ∩ I	12	$n(B) - n(B \cap C' \cap I')$ $\qquad - n(B \cap C)$ $= 77 - 24 - 41 = 12$
C and I and not B	B' ∩ C ∩ I	6	$n(C) - n(C \cap B' \cap I')$ $\qquad - n(B \cap C)$ $= 65 - 18 - 41 = 6$
I Only	B' ∩ C' ∩ I	0	$n(I) - n(B \cap I \cap C')$ $\qquad - n(B \cap C \cap I)$ $\qquad - n(C \cap I \cap B')$ $= 27 - 12 - 9 - 6 = 0$

Construct a Venn diagram from the table.

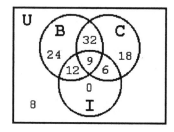

$n(U) = 8 + 9 + 24 + 18 + 32 + 12 + 6 + 0 = 109$

13. Continued

   a) n(infrared film only) = n(B′ ∩ C′ ∩ I) = 0

      In percent, $\dfrac{n(B′ \cap C′ \cap I)}{n(U)} = \dfrac{0}{109} = 0.0 = 0\%$

   b) T = {members|the member used at least two types of film}

      $$n(T) = n(B \cap C \cap I) + n(B \cap C \cap I′) + n(B \cap I \cap C′)$$
      $$\qquad + n(C \cap I \cap B′)$$
      $$= 9 + 32 + 12 + 6$$
      $$= 59 \text{ members used at least two types of film}$$

      In percent, $\dfrac{n(T)}{n(U)} = \dfrac{59}{109} = 0.541284 \approx 54\%.$

17. U = {persons|the person is a pet owner)
    F = {persons|the person owns fish)
    B = {persons|the person owns a bird)
    C = {persons|the person owns a cat)
    D = {persons|the person owns a dog)

   n(U) = 136,
   n(F) = 49,
   n(B) = 55,
   n(C) = 50,
   n(D) = 68,
   n(all) = n(F ∩ B ∩ C ∩ D) = 2,
   n(F only) = n(F ∩ B′ ∩ C′ ∩ D′) = 11,
   n(B only) = n(F′ ∩ B ∩ C′ ∩ D′) = 14,
   n(F and B) = n(F ∩ B) = 10,
   n(F and C) = n(F ∩ C) = 21,
   n(B and D) = n(B ∩ D) = 26,
   n(C and D) = n(C ∩ D) = 27,
   n(F and B and C, no D) = n(F ∩ B ∩ C ∩ D′) = 3,
   n(F and B and D, no C) = n(F ∩ B ∩ C′ ∩ D) = 1,
   n(F and C and D, no B) = n(F ∩ B′ ∩ C ∩ D) = 9,
   n(B and C and D, no F) = n(F′ ∩ B ∩ C ∩ D) = 10

Exercise 2.2

17. Continued

Region	Set	n(Set)	Calculation
All	F ∩ B ∩ C ∩ D	2	Given
F and B and C, no D	F ∩ B ∩ C ∩ D′	3	Given
F and B and D, no C	F ∩ B ∩ C′ ∩ D	1	Given
F and C and D, no B	F ∩ B′ ∩ C ∩ D	9	Given
B and C and D, no F	F′ ∩ B ∩ C ∩ D	10	Given

Place the above regions on the Venn diagram as follows:

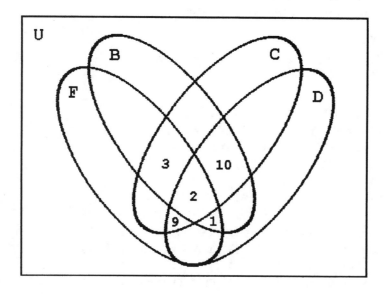

17. **Continued**

Region	Set	n(Set)	Calculation
F and B, not C or D	F ∩ B ∩ C′ ∩ D′	4	Using n(F ∩ B) and figure above, subtract to find n(Set). 10 − (3 + 2 + 1) = 10 − 6 = 4
F and C, not B or D	F ∩ B′ ∩ C ∩ D′	7	Using n(F ∩ C) and figure above, subtract to find n(Set). 21 − (3 + 2 + 9) = 21 − 14 = 7
B and D, not F or C	F′ ∩ B ∩ C′ ∩ D	13	Using n(B ∩ D) and figure above, subtract to find n(Set). 26 − (10 + 2 + 1) = 26 − 13 = 13
C and D, not F or B	F′ ∩ B′ ∩ C ∩ D	6	Using n(C ∩ D) and figure above, subtract to find n(Set). 27 − (10 + 2 + 9) = 27 − 21 = 6

Place the above regions on the Venn diagram as follows:

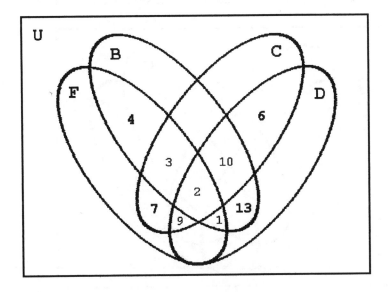

17. Continued

Region	Set	n(Set)	Calculation
F only	F ∩ B′ ∩ C′ ∩ D′	11	Given
F and D, not B or C	F ∩ B′ ∩ C′ ∩ D	12	Using n(F) and figure above, subtract to find n(Set). 49 − (11 + 4 + 7 + 3 + 9 + 2 + 1) = 49 − 37 = 12
B only	F′ ∩ B ∩ C′ ∩ D′	14	Given
B and C, not F or D	F′ ∩ B ∩ C ∩ D′	8	Using n(B) and figure above, subtract to find n(Set). 55 − (4 + 14 + 3 + 10 + 2 + 1 + 13) = 55 − 47 = 8
C only	F′ ∩′ B ∩ C ∩ D′	5	Using n(C) and figure above, subtract to find n(Set). 50 − (7 + 3 + 8 + 9 + 2 + 10 + 6) = 50 − 45 = 5
D only	F′ ∩ B′ ∩ C′ ∩ D	15	Using n(D) and figure above, subtract to find n(Set). 68 − (9 + 2 + 10 + 6 + 12 + 1 + 13) = 68 − 53 = 15

Place the above regions on the Venn diagram as follows:

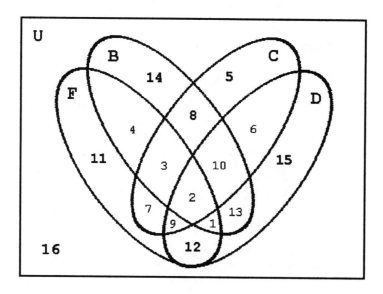

17. **Continued**

$n(U) - n(F) - n(B \cap F') - n(F' \cap B' \cap C \cap D) - n(C \text{ only})$
$- n(D \text{ only})$
$$= 136 - 49 - (14 + 8 + 10 + 13) - 6 - 5 - 15$$
$$= 136 - 49 - 45 - 26 = 136 - 120$$
$$= 16 \text{ pet owners who have no fish, no}$$
$$\text{cats, and no dogs}$$

21. $A' = \{1,3,6,7,8\}, \; B = \{1,2,7,8,9\}$

$(A \cap B')' = A' \cup (B')' = A' \cup B$
$$= \{1,2,3,6,7,8,9\}$$

25. Fill in the Venn diagram as shown.

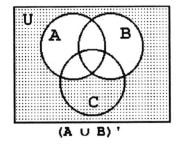 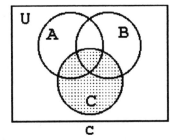 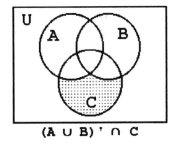

$(A \cup B)'$ C $(A \cup B)' \cap C$

29. <u>Region</u>   <u>Description</u>

     I   Neither A nor B, $(A \cup B)'$
     II   A and not B, $A \cap B'$
     III   Both A and B, $A \cap B$
     IV   B and not A, $A' \cap B$

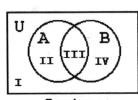

Regions

The set $(A \cap B)$ consists of all elements in region III. Therefore, the complement $(A \cap B)'$ consists of regions I, II and IV.

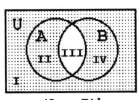

$(A \cap B)'$

43

29.   Continued

Now, A′ consists of all elements in regions
I and IV.

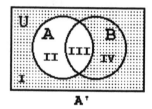

And B′ consists of all elements in regions I
and II.

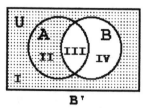

So, the union of A′ and B′ (A′ ∪ B′) would
consist of all elements in regions I, II
and IV.

Therefore, (A ∩ B)′ has the same shaded
area as (A′ ∪ B′) and we can conclude that
(A ∩ B)′ = A′ ∪ B′.

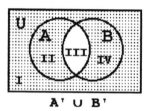

33.   The correct answer is (d). The first restriction excludes (a).
The second restriction excludes (b) and (e). The third
restriction excludes (c).

37.   The correct answer is (b). If the meeting is not at Frank's
house, it must be at Betty's or Delores's house. By the last
restriction, Angela and Carmine attend. The third restriction
states that Angela cannot attend any meetings at Delores's house
so the meeting must be held at Betty's house. Delores cannot
attend if the meeting is held at Betty's house so (e) is
excluded. Betty, Angela and Carmine attend so (a) is excluded.
No information is given whether Ed and/or Grant attend so that
excludes (c) and (d).

Exercise 2.3

1. There are two possible outcomes (heads or tails) on each coin.

   a) Create a box for each coin and enter possible outcomes and then multiply.

   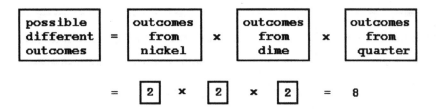

   b) Tree diagram:

   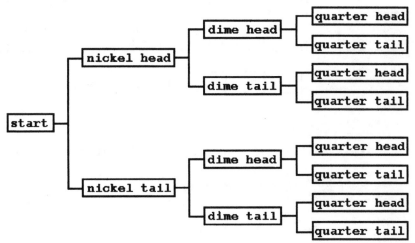

5.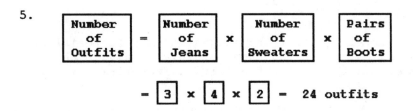

Exercise 2.3

9.

Number of Ski Equip. Packages	=	Lines of Snow Skis	x	Types of Bindings	x	Types of Boots	x	Types of Poles

$$= \boxed{14} \times \boxed{7} \times \boxed{9} \times \boxed{3}$$

$$= 2646 \text{ ski equipment packages}$$

13.

Number of Social Security Numbers	=	Ten choices for each of nine digits

$$= 10 \cdot 10 \cdot 10 \cdot 10 \cdot 10 \cdot 10 \cdot 10 \cdot 10 \cdot 10 = 10^9$$
$$= 1,000,000,000$$
$$= 1 \text{ billion possible Social Security numbers}$$

17.

Number of telephone area codes	=	first digit (not 0 or 1)	x	second digit (only 0 or 1)	x	third digit (not 0 or 1)

$$= \boxed{8} \times \boxed{2} \times \boxed{8} = 128 \text{ area codes}$$

21. $4! = 4 \cdot 3 \cdot 2 \cdot 1 = 24$

25. $20! = 20 \cdot 19 \cdot 18 \cdot 17 \cdot \ldots \cdot 2 \cdot 1 = 2.4329 \times 10^{18}$ (Use calculator)

29. a) $\dfrac{6!}{4!} = \dfrac{6 \cdot 5 \cdot 4 \cdot 3 \cdot 2 \cdot 1}{4 \cdot 3 \cdot 2 \cdot 1} = 6 \cdot 5 = 30$

   b) $\dfrac{6!}{2!} = \dfrac{6 \cdot 5 \cdot 4 \cdot 3 \cdot 2 \cdot 1}{2 \cdot 1} = 6 \cdot 5 \cdot 4 \cdot 3 = 360$

33. $\dfrac{8!}{4!4!} = \dfrac{8 \cdot 7 \cdot 6 \cdot 5 \cdot 4!}{4! \cdot 4 \cdot 3 \cdot 2 \cdot 1} = \dfrac{8 \cdot 7 \cdot 6 \cdot 5}{4 \cdot 3 \cdot 2 \cdot 1} = \dfrac{1680}{24} = 70$

When dividing factorials, cancel the largest common factorial in the denominator and numerator and then multiply or divide the remaining numbers. In this case, the largest factorial was 4!.

37. $n = 16, \ r = 14$

$$\frac{n!}{(n-r)!} = \frac{16!}{(16-14)!} = \frac{16!}{2!} = \frac{16 \cdot 15 \cdot 14 \cdot \ldots \cdot 3 \cdot 2!}{2!} = \frac{16 \cdot 15 \cdot 14 \cdot \ldots \cdot 3}{1}$$

$$= 1.04613949 \times 10^{13}$$

41. $n = 7, \ r = 3$

$$\frac{n!}{(n-r)!r!} = \frac{7!}{(7-3)!3!} = \frac{7!}{4!3!} = \frac{7 \cdot 6 \cdot 5 \cdot 4!}{4! \cdot 3 \cdot 2 \cdot 1} = \frac{7 \cdot 6 \cdot 5}{3 \cdot 2 \cdot 1} = \frac{210}{6} = 35$$

45. Omitted.

49. The correct choice is (a). The gray car is parked in space #2.
     The pink car is parked in #3. The black car and yellow car must
     be parked next to each other so they would have to be parked in
     two of the spaces #4, #5 and #6. Since the gray car and the
     white car must have at least one space between them, the white
     car must also be parked in one of the spaces #4, #5, or #6.
     Leaving only space #1 that can be empty.

Exercise 2.4

1. a)
    $$_nP_r = \frac{n!}{(n-r)!}, \quad \text{permutation where } n = 7, \ r = 3$$

    $$_7P_3 = \frac{7!}{(7-3)!} = \frac{7!}{4!} = \frac{7 \cdot 6 \cdot 5 \cdot 4!}{4!} = 7 \cdot 6 \cdot 5 = 210$$

    b)
    $$_nC_r = \frac{n!}{(n-r)!r!}, \quad \text{combination where } n = 7, \ r = 3$$

    $$_7C_3 = \frac{7!}{(7-3)!3!} = \frac{7 \cdot 6 \cdot 5 \cdot 4!}{4!3!} = \frac{7 \cdot 6 \cdot 5}{3 \cdot 2 \cdot 1} = \frac{210}{6} = 35$$

5. a)
    $$_nP_r = \frac{n!}{(n-r)!}, \quad \text{permutation where } n = 14, \ r = 1$$

    $$_{14}P_1 = \frac{14!}{(14-1)!} = \frac{14!}{13!} = \frac{14 \cdot 13!}{13!} = 14$$

    b)
    $$_nC_r = \frac{n!}{(n-r)!r!}, \quad \text{combination where } n = 14, \ r = 1$$

    $$_{14}C_1 = \frac{14!}{(14-1)!1!} = \frac{14 \cdot 13!}{13!1!} = \frac{14}{1} = 14$$

Exercise 2.4

9.   a) $$_nP_r = \frac{n!}{(n - r)!}, \quad \text{permutation where } n = x, \ r = x - 1$$

$$_xP_{x-1} = \frac{x!}{[x - (x - 1)]!} = \frac{x!}{[x - x + 1]!} = \frac{x!}{1!} = x!$$

   b) $$_nC_r = \frac{n!}{(n - r)!r!}, \quad \text{combination where } n = x, \ r = x - 1$$

$$_xC_{x-1} = \frac{x!}{[x - (x - 1)]!(x - 1)!} = \frac{x \cdot (x - 1)!}{1![x - 1]!} = \frac{x}{1} = x$$

13. a) $$_nP_r = \frac{n!}{(n - r)!}, \quad \text{permutation where } n = 3, \ r = 2$$

$$_3P_2 = \frac{3!}{(3 - 2)!} = \frac{3!}{1!} = \frac{3 \cdot 2 \cdot 1}{1} = 6$$

   b)   {a,b}, {a,c}, {b,a}, {b,c}, {c,a}, {c,b}

17. 12 students, each to give a presentation.

   a)   The order **is** important so use a **permutation**

$$_nP_r = \frac{n!}{(n - r)!}, \quad \text{permutation where } n = 12, \ r = 12$$

$$_{12}P_{12} = \frac{12!}{(12 - 12)!} = \frac{12!}{0!} = \frac{12 \cdot 11 \cdot \ldots \cdot 1}{1} = 479,001,600$$

   b)   There is only one way in which the presentation order will be alphabetical.

21. There are 14 teams, each team playing every other team once. The order in which they play **is not** important so **combinations** are used.

$$_nC_r = \frac{n!}{(n - r)!r!}, \quad \text{combination where } n = 14, \ r = 2$$

$$_{14}C_1 = \frac{14!}{(14 - 2)! \cdot 2!} = \frac{14 \cdot 13 \cdot 12!}{12! \cdot 2 \cdot 1} = \frac{14 \cdot 13}{2 \cdot 1} = \frac{182}{2} = 91$$

25. Select a four-person committee from 8 women and 6 men. The Fundamental Principle of Counting tells us to multiply the number from each category (women and men) together. In a committee the order is not important so combinations will be used.

$$_nC_r = \frac{n!}{(n - r)!r!}$$

25. Continued

a)   Must have 2 women and 2 men.

Number of Possible Committees	=	Select 2 women from 8 (n = 8, r = 2)	×	Select 2 men from 6 (n = 6, r = 2)

$$= \left({}_8C_2\right) \bullet \left({}_6C_2\right)$$

$$= \frac{8!}{(8-2)!2!} \bullet \frac{6!}{(6-2)!2!}$$

$$= \frac{8!}{6!2!} \bullet \frac{6!}{4!2!} = \frac{8 \bullet 7}{2 \bullet 1} \bullet \frac{6 \bullet 5}{2 \bullet 1} = \frac{56}{2} \bullet \frac{30}{2} = 28 \bullet 15 = 420$$

b)   For any mixture we add women and men together to obtain 4 possible members. The order is not important so we will use combinations where n = 14 and r = 4.

$$_{14}C_4 = \frac{14!}{(14-4)!4!} = \frac{14 \bullet 13 \bullet 12 \bullet 11 \bullet 10!}{10! \bullet 4 \bullet 3 \bullet 2 \bullet 1} = \frac{14 \bullet 13 \bullet 12 \bullet 11}{4 \bullet 3 \bullet 2 \bullet 1}$$

$$= \frac{24,024}{24} = 1001$$

c)   To have a majority of women there must be more women than men. It is possible to meet this condition either by having 3 women and 1 man **or** by having 4 women and no men. Calculate the possible outcomes from each case and then **add** the results together.

Case 1:   3 women and 1 man

Number of Possible Committees	=	Select 3 women from 8 (n = 8, r = 3)	×	Select 1 man from 6 (n = 6, r = 1)

$$= \left({}_8C_3\right) \bullet \left({}_6C_1\right)$$

$$= \frac{8!}{(8-3)!3!} \bullet \frac{6!}{(6-1)!1!}$$

$$= \frac{8!}{5!3!} \bullet \frac{6!}{5!1!} = \frac{8 \bullet 7 \bullet 6 \bullet 5!}{5! \bullet 3 \bullet 2 \bullet 1} \bullet \frac{6 \bullet 5!}{5! \bullet 1} = \frac{336}{6} \bullet \frac{6}{1} = 336$$

Exercise 2.4

25.c) Continued

Case 2: 4 women and no men

Number of Possible Committees	=	Select 4 women from 8 ($n = 8$, $r = 4$)	x	Select no men from 6 ($n = 6$, $r = 0$)

$$= (_8C_4) \cdot (_6C_0)$$

$$= \frac{8!}{(8 - 4)!4!} \cdot \frac{6!}{(6 - 0)!0!}$$

$$= \frac{8!}{4!4!} \cdot \frac{6!}{6!1} = \frac{8 \cdot 7 \cdot 6 \cdot 5 \cdot 4!}{4 \cdot 3 \cdot 2 \cdot 1 \cdot 4!} \cdot \frac{1}{1} = \frac{1680}{24} \cdot \frac{1}{1} = 70$$

Adding Case 1 and Case 2 together, the total possible committees will be:

Case 1 + Case 2 = 336 + 70 = 406

29. Select a five-card poker hand from a deck of cards consisting of four suits with 13 different cards per suit. The Fundamental Principle of Counting tells us to multiply the number from each category (suits and cards) together. In a poker hand order is not important so combinations will be used.

$$_nC_r = \frac{n!}{(n - r)!r!}, \quad \text{combination where n and r vary}$$

a) Must have 3 aces out of 4 possible and then have 2 cards from the remaining 48 cards.

Number of Possible Poker Hands	=	Select 3 suits from 4 ($n = 4$, $r = 3$)	x	Select 2 cards from 48 ($n = 48$, $r = 2$)

$$= (_4C_3) \cdot (_{48}C_2)$$

$$= \frac{4!}{(4 - 3)!3!} \cdot \frac{48!}{(48 - 2)!2!}$$

$$= \frac{4 \cdot 3!}{1!3!} \cdot \frac{48 \cdot 47 \cdot 46!}{46! \cdot 2 \cdot 1} = \frac{4}{1} \cdot \frac{48 \cdot 47}{2 \cdot 1} = \frac{4}{1} \cdot \frac{2256}{2}$$

$$= (4)(1128) = 4512 \quad \text{possible hands with exactly three aces}$$

29. Continued

b) Must have 1 card out of 13 and then have 3 different suits of that card and then any 2 cards from the remaining 48. There are 13 ways to get the first card.

Number of Possible Poker Hands	=	13 ways to select 3 cards from 4 (n = 4, r = 3)	x	Select 2 cards from 48 (n = 48, r = 2)

$$= 13(_4C_3) \cdot (_{48}C_2)$$

$$= 13 \cdot \frac{4!}{(4-3)!3!} \cdot \frac{48!}{(48-2)!2!}$$

$$= 13 \cdot \frac{4 \cdot 3!}{1!3!} \cdot \frac{48 \cdot 47 \cdot 46!}{46! \cdot 2 \cdot 1} = 13 \cdot \frac{4}{1} \cdot \frac{48 \cdot 47}{2 \cdot 1} = 13 \cdot \frac{4}{1} \cdot \frac{2256}{2}$$

$$= (13)(4)(1128) = 58,656 \text{ possible hands with three-of-a-kind}$$

33. Choose 6 numbers out of 53. The order **is not** important so use a **combination**.

$$_nC_r = \frac{n!}{(n-r)!r!}, \quad \text{combination where } n = 53, r = 6$$

$$_{53}C_6 = \frac{53!}{(53-6)!6!} = \frac{53 \cdot 52 \cdot 51 \cdot 50 \cdot 49 \cdot 48 \cdot 47!}{47! \cdot 6 \cdot 5 \cdot 4 \cdot 3 \cdot 2 \cdot 1}$$

$$= \frac{53 \cdot 52 \cdot 51 \cdot 50 \cdot 49 \cdot 48}{6 \cdot 5 \cdot 4 \cdot 3 \cdot 2 \cdot 1} = \frac{1.65293856 \times 10^{10}}{720}$$

$$= 22,957,480 \text{ possible lottery tickets}$$

37. The 5/36 lottery would be easier to win because there are fewer combinations of numbers that can be chosen.

41. Pascal's Triangle is Figure 2.28.
    a) If the first row is counted as the 0th row, then $_4C_2$ would be found in the 4th row.

    b) Counting the first row as the 0th row, $_nC_r$ would be found in the nth row.

    c) No, the second number in the fourth row is not $_4C_2$.

    d) Yes, the third number in the fourth row is $_4C_2$.

    e) The location of $_nC_r$ would be in the nth row (counting the first row as the 0th row) and the $(r + 1)$ number.

Exercise 2.4

45. Team B must play team D second, so (c) is the correct choice. In the second set of games, B cannot play

        A because A is playing F (first condition),
        C because B plays C in the third game (second
           condition),
        E because B played E in the first game (second
           condition), or
        F because F is playing A (first condition).

Exercise 2.5

1. S is equivalent to C (S ~ C) because they both contain 4 elements, $n(S) = n(C) = 4$.

   Possible one-to-one correspondence (capital city $\leftrightarrow$ state):
       Sacramento $\leftrightarrow$ California
       Lansing $\leftrightarrow$ Michigan
       Richmond $\leftrightarrow$ Virginia
       Topeka $\leftrightarrow$ Kansas

5. C contains 22 multiples of 3, so $n(C) = 22$ and D contains 22 multiples of 4, so $n(D) = 22$. Therefore, C and D have the same cardinal number, C ~ D.

   Possible one-to-one correspondence ($3n \leftrightarrow 4n$):
   $$C = \{3,\ 6,\ 9,\ 12,\ \ldots,\ 63,\ 66\}$$
   $$\updownarrow\ \updownarrow\ \updownarrow\ \updownarrow\ \ \updownarrow\ \ \updownarrow\ \updownarrow$$
   $$D = \{4,\ 8,\ 12,\ 16,\ \ldots,\ 84,\ 88\}$$

9. A is the set of odd numbers from 1 through 123 which can be represented by $2n - 1$. The last number, 123, would be the 62nd number in the set, $2(62) - 1 = 124 - 1 = 123$. So $n(A) = 62$.

   B is the set of odd numbers from 125 through 247 which can be represented by $2n + 123$. The last number, 247 would be the 62nd number in the set, $2(62) + 123 = 124 + 123 = 247$. So $n(B) = 62$.

   The cardinal number of A and B are the same, so A ~ B.

   Possible one-to-one correspondence ($2n - 1 \leftrightarrow 2n + 123$):
   $$A = \{\ 1,\ \ \ 3,\ \ \ 5,\ \ \ldots,\ \ \ 2n - 1,\ \ \ \ldots,\ 121,\ 123\}$$
   $$\updownarrow\ \ \ \updownarrow\ \ \ \updownarrow\ \ \ \updownarrow\ \ \ \ \ \ \updownarrow\ \ \ \ \ \ \updownarrow\ \ \ \ \updownarrow$$
   $$B = \{125,\ 127,\ 129,\ \ldots,\ \ 2n + 123,\ \ldots,\ 245,\ 247\}$$

13. a) The elements of N and T can be paired up as follows:

$$N = \{1, \quad 2, \quad 3, \quad 4, \quad ..., \quad n, \quad n + 1, \quad ...\}$$
$$\updownarrow \quad \updownarrow \quad \updownarrow \quad \updownarrow \quad \updownarrow \quad \updownarrow \quad \updownarrow \qquad \updownarrow$$
$$T = \{3, \quad 6, \quad 9, \quad 12, \quad ..., \quad 3n, \quad 3(n + 1), \quad ...\}$$

Any natural number, n ∈ N, corresponds with the multiple of three, 3n ∈ T. There exists a one-to-one correspondence between the elements of N and T, therefore the sets are equivalent; that is, t ~ N.

b) $936 = 3n \in T$, so $n = \dfrac{936}{3} = 312 \in N$

c) $x = 3n \in T$, so $n = \dfrac{x}{3} \in N$

d) $936 = n \in N$, so $3n = 3(936) = 2808 \in T$

e) n ∈ N, so 3n ∈ T

17. Let A'B = line segment [0,1] and AC = line segment [0,3]. Line segment AC is three times as long as line segment A'B. Draw A'B above AC and extend line segments AA' and CB so that they meet at point D.

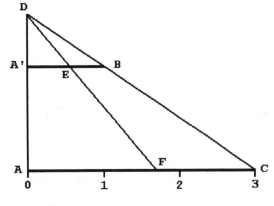

Any point E on A'B can be paired up with the unique point F on AC formed by the intersection of lines DE and AC. Conversely, any point F on segment AC can be paired up with the unique point E on A'B formed by the intersection of lines DF and A'B.

Therefore, a one-to-one correspondence exists between the two segments and the interval [0,1] contains exactly the same number of points as the interval [0,3]. Thus, the intervals are equivalent.

21. Draw the circle of radius 1 cm
    inside the circle of radius 5 cm
    so that they have the same center
    C.

    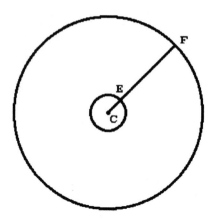

    Any point E on the smaller circle
    can be paired with the unique
    point F on the larger circle by
    drawing a line from the center of
    the circle through E. Conversely,
    any point F on the larger circle
    can be paired with the unique
    point E on the inner circle by
    drawing a line from the center of
    the circle through F.

    Therefore, a one-to-one correspondence exists between the two
    circles and they are equivalent.

25. Omitted.

Chapter 2 Review

1. Omitted.

5. a)   n(U) = 61, n(A) = 32, n(B) = 26, n(A ∪ B) = 40

   $$n(A ∪ B) = n(A) + n(B) - n(A ∩ B)$$
   $$40 = 32 + 26 - n(A ∩ B)$$
   $$40 = 58 - n(A ∩ B)$$

   n(A ∩ B) = 58 - 40 = 18

   b)   n(A only) = n(A ∩ B') = 32 - 18 = 14

   n(B only) = n(B ∩ A') = 26 - 18 = 8

   n(None) = n(A ∪ B)' = 61 - 40 = 21

   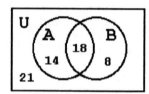

9. a) Create a box for each event, enter each possible outcome and then multiply.

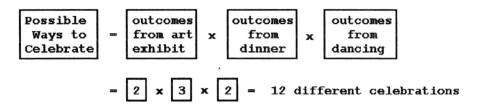

$$= \boxed{2} \times \boxed{3} \times \boxed{2} = 12 \text{ different celebrations}$$

b) Tree diagram:

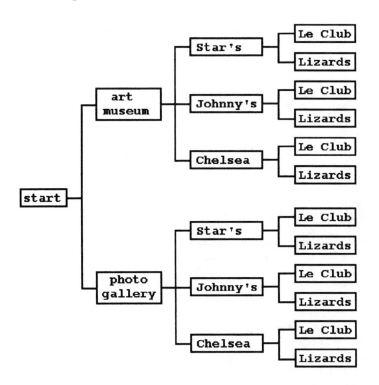

13. 11 items, 3 at a time.

a) The order is **not** important so use a **combination**

$$_nC_r = \frac{n!}{(n-r)!\,r!}, \quad \text{combination where } n = 11, \ r = 3$$

$$_{11}C_3 = \frac{11!}{(11-3)!\,3!} = \frac{11 \cdot 10 \cdot 9 \cdot 8!}{8! \cdot 3 \cdot 2 \cdot 1} = \frac{11 \cdot 10 \cdot 9}{3 \cdot 2 \cdot 1} = \frac{990}{6} = 165$$

13. Continued

   b)   The order **is** important so use a **permutation**

   $$_nP_r = \frac{n!}{(n-r)!},\quad \text{permutation where } n = 11,\ r = 3$$

   $$_{11}P_3 = \frac{11!}{(11-3)!} = \frac{11\cdot10\cdot9\cdot8!}{8!} = \frac{11\cdot10\cdot9}{1} = 990$$

17. 10 teams, 3 end-of-the-season rankings. The order **is** important so use a **permutation**.

   $$_nP_r = \frac{n!}{(n-r)!},\quad \text{permutation where } n = 10,\ r = 3$$

   $$_{10}P_3 = \frac{10!}{(10-3)!} = \frac{10\cdot9\cdot8\cdot7!}{7!} = \frac{10\cdot9\cdot8}{1} = 720$$

21. Omitted.

25. C is the set of odd numbers from 3 through 901 which can be represented by $2n + 1$. The last number, 901, would be the 450th number in the set, $2(450) + 1 = 901$. So $n(C) = 450$.

   D is the set of even numbers from 2 through 900 which can be represented by $2n$. The last number, 900 would be the 450th number in the set, $2(450) = 900$. So $n(D) = 450$.

   The cardinal number of C and D are the same, so $C \sim D$.

   Possible one-to-one correspondence $(2n + 1 \leftrightarrow 2n)$:
   $$C = \{3,\ 5,\ 7,\ \ldots,\ 2n + 1,\ \ldots,\ 899,\ 901\}$$
   $$\updownarrow\ \updownarrow\ \updownarrow\ \updownarrow\qquad \updownarrow\qquad \updownarrow\qquad \updownarrow\quad \updownarrow$$
   $$D = \{2,\ 4,\ 6,\ \ldots,\quad 2n,\quad \ldots,\ 898,\ 900\}$$

29. Let A'B = line segment [0,1] and AC = line segment [0,π]. Line segment AC is pi times as long as line segment A'B. Draw A'B above AC and extend line segments AA' and CB so that they meet at point D.

   Any point E on A'B can be paired up with the unique point F on AC formed by the intersection of lines

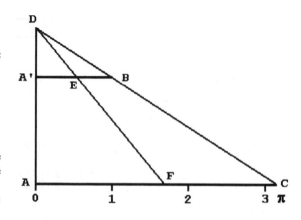

DE and AC. Conversely, any point F on segment AC can be paired up with the unique point E on A'B formed by the intersection of lines DF and A'B.

Therefore, a one-to-one correspondence exists between the two segments and the interval [0,1] contains exactly the same number of points as the interval [0,π]. Thus, the intervals are equivalent.

3   Probability

Exercise 3.1

1.   Individual solutions will vary depending on what happens when the
      single die is rolled. The following table records the results of
      a trial of this experiment.

Result (Number of 6s rolled in four rolls of die)	Number of times in 10 trials that result occurred
0	3
1	4
2	3
3	0
4	0

Betting that you could roll at least one six, you would have won
7 times ($70) and lost three times ($30). Thus, you would have
won $40 (70 − 30 = 40).

5.   You bet $5 on a 17-20 split. The house odds are 17 to 1.
      a)   You win $85. (5•17 = 85)
      b)   You win $85. (5•17 = 85)
      c)   You lose $5.

9.   You bet $10 on 31-32-33-34-35-36 line. The house odds are 5
      to 1.
      a)   You lose $10.
      b)   You win $50. (10•5 = 50)

13.   You bet $50 on the odd numbers. The house odds are 1 to 1.
      a)   You lose $50.
      b)   You win $50.

17. You bet $30 on the 1-2 split (house odds are 17 to 1), and also
    bet $15 on the even numbers (house odds are 1 to 1).
    a)   You win $495.   (30•17 – 15•1 = 510 – 15 = 495)
    b)   You win $525.   (30•17 + 15•1 = 510 + 15 = 525)
    c)   You lose $45.   (–30 – 15 = –45)
    d)   You lose $15.   (–30 + 15•1 = –30 + 15 = 15)

21. Since the house odds are 2 to 1 .you must bet $1000/2 =$500.

25. a)   There are 12 face cards in a deck of cards. A jack, queen
         and king from each of 4 suits.

    b)   There are 52 cards in a deck of cards. The fraction that
         are face cards is 12/52 = 3/13.

29. Omitted.

33. Omitted.

37. Omitted.

Exercise 3.2

1. The experiment is the picking of a jellybean.

5. Event A = picking a red or yellow jelly bean
    $n(S) = 12 + 8 + 10 + 5 = 35$, $n(A) = 8 + 10 = 18$

    $$p(A) = \frac{n(A)}{n(S)} = \frac{18}{35}$$

    This means that 18 out of every 35 possible outcomes are a
    success.

9. Event B = picking a white jelly bean
    $n(S) = 12 + 8 + 10 + 5 = 35$, $n(B) = 0$

    $$p(B) = \frac{n(B)}{n(S)} = \frac{0}{35} = 0$$

    This means that there is no possible ways of a successful
    outcome.

Exercise 3.2

13. Event A = picking a red or yellow jelly bean
    n(S) = 12 + 8 + 10 + 5 = 35, n(A) = 8 + 10 = 18
    n(A') = n(S) – n(A) = 35 – 18 = 17

    o(B) = n(A):N(A') = 18:17

    This means that there are 18 successful outcomes for every 17 possible failures.

17. E = drawing a black card
    n(S) = 52, n(E) = 26, n(E') = n(S) – n(E) = 52 – 26 = 26

    a) $p(E) = \dfrac{n(E)}{n(S)} = \dfrac{26}{52} = \dfrac{1}{2}$

    This means that one out of every two possible outcomes is a success (that is, a black card).

    b) o(E) = n(E):n(E') = 26:26 = 1:1

    This means that there is one possible success for every one possible failure.

21. E = drawing a queen of spades
    n(S) = 52, n(E) = 1, n(E') = n(S) – n(E) = 52 – 1 = 51

    a) $p(E) = \dfrac{n(E)}{n(S)} = \dfrac{1}{52}$

    This means that there is one possible success for every 52 possible outcomes.

    b) o(E) = n(E):n(E') = 1:51

    This means that there is one possible success for every 51 possible failures.

25. E = drawing a card above a 4 = {5, 6, 7, 8, 9, 10, J, Q, K, A}
    E' = {2, 3, 4}
    n(S) = 52, n(E) = 10(4 suits) = 40, n(E') = 3(4 suits) = 12

    a) $p(E) = \dfrac{n(E)}{n(S)} = \dfrac{40}{52} = \dfrac{10}{13}$

    This means that there are ten possible successes for every 13 possible outcomes.

25. Continued

b) $o(E) = n(E):n(E') = 40:12 = 10:3$

This means that there are ten possible successes for every 3 possible failures.

29. E = ball lands on a single number
$n(S) = 38$, $n(E) = 1$, $n(E') = n(S) - n(E) = 38 - 1 = 37$

a) $p(E) = \dfrac{n(E)}{n(S)} = \dfrac{1}{38}$

This means that there is one possible success for every 38 possible outcomes.

b) $o(E) = n(E):n(E') = 1:37$

This means that there is one possible success for every 37 possible failures.

33. E = ball lands on one of five numbers
$n(S) = 38$, $n(E) = 5$, $n(E') = n(S) - n(E) = 38 - 5 = 33$

a) $p(E) = \dfrac{n(E)}{n(S)} = \dfrac{5}{38}$

This means that there are 5 possible successes for every 38 possible outcomes.

b) $o(E) = n(E):n(E') = 5:33$

This means that for every five possible successes there are 33 possible failures.

37. E = ball lands on an even number
$n(S) = 38$, $n(E) = 18$, $n(E') = n(S) - n(E) = 38 - 18 = 20$

a) $p(E) = \dfrac{n(E)}{n(S)} = \dfrac{18}{38} = \dfrac{9}{19}$

This means that there are 9 possible successes for every 19 possible outcomes.

b) $o(E) = n(E):n(E') = 18:20 = 9:10$

This means that for every 9 possible successes there are 10 possible failures.

Exercise 3.2

41. $o(E) = n(E):n(E') = 3:2$, so $n(E) = 3$, $n(E') = 2$.
$n(S) = N(E) + n(E') = 3 + 2 = 5$.

$p(E) = \dfrac{n(E)}{n(S)} = \dfrac{3}{5}$

45. E = ball lands on an odd number
$n(S) = 38$, $n(E) = 18$, $n(E') = n(S) - n(E) = 38 - 18 = 20$

a) $p(E) = \dfrac{n(E)}{n(S)} = \dfrac{18}{38} = \dfrac{9}{19} = \dfrac{a}{b}$

This means that there are $a = 9$ possible successes for every $b = 19$ possible outcomes.

b) $o(E) = a:(b - a) = 9:(19 - 9) = 9:10$

This means that for every $a = 9$ possible successes there are $(b - a) = 10$ possible failures.

49. a) $S = \{(b,b,b),\ (b,b,g),\ (b,g,b),\ (b,g,g),\ (g,b,b),$
$(g,b,g),\ (g,g,b),\ (g,g,g)\}$

b) $E = \{(b,g,g),\ (g,b,g),\ (g,g,b)\}$

c) $F = \{(b,g,g),\ (g,b,g),\ (g,g,b),\ (g,g,g)\}$

d) $G = \{(g,g,g)\}$

For e) through j): $n(S) = 8$, $n(E) = 3$, $n(F) = 4$, $n(G) = 1$

e) $p(E) = \dfrac{n(E)}{n(S)} = \dfrac{3}{8}$

f) $p(F) = \dfrac{n(F)}{n(S)} = \dfrac{4}{8} = \dfrac{1}{2}$

g) $p(G) = \dfrac{n(G)}{n(S)} = \dfrac{1}{8}$

h) $o(E) = n(E):n(E') = n(E):[n(S) - n(E)] = 3:(8 - 3) = 3:5$

i) $o(F) = n(F):n(F') = n(F):[n(S) - n(F)] = 4:(8 - 4) = 4:4$
$= 1:1$

j) $o(G) = n(G):n(G') = n(G):[n(S) - n(G)] = 1:(8 - 1) = 1:7$

53. The couple would be more likely to have children of different sexes because there are only 2 ways of having the same sex, {b, b, b} or {g, g, g}, while there are six ways of having three children of opposite sexes. (See Exercise 49.)

57. Pink snapdragons have one red gene (r) and one white gene (w). The Punnett square would look like the following:

	r	w	
r	(r, r)	(w, r)	← offspring
w	(r, w)	(w, w)	← offspring

↑ first parent's genes (r column header) — ← first parent's genes

↑
second parent's genes

a) R = {(r, r)} (red offspring)

$$p(R) = \frac{n(R)}{n(S)} = \frac{1}{4}$$

b) W = {(w, w)} (white offspring)

$$p(W) = \frac{n(W)}{n(S)} = \frac{1}{4}$$

c) P = {(r, w), (w, r)} (pink offspring)

$$p(P) = \frac{n(P)}{n(S)} = \frac{2}{4} = \frac{1}{2}$$

61. One parent is a carrier of Tay-Sachs disease and one parent is not. Let T denote the disease-free gene and t denote the recessive Tay-Sachs disease gene. The Punnett square would look like the following:

	T	t	
T	(T, T)	(t, T)	← offspring
T	(T, T)	(t, T)	← offspring

← carrier parent's genes

↑
non-carrier parent's genes

a) D = {(t, t)} = Ø (child has the disease)

$$p(D) = \frac{n(D)}{n(S)} = \frac{0}{4} = 0$$

61. Continued

    b)   C = { (t,T), (t,T) }   (child is a carrier)

$$p(C) = \frac{n(C)}{n(S)} = \frac{2}{4} = \frac{1}{2}$$

    c)   H = { (T,T), (T,T), (t,T), (t,T) }   (child is healthy---
                                                                        includes carriers)

$$p(H) = \frac{n(H)}{n(S)} = \frac{4}{4} = 1$$

65. Omitted.

69. Omitted.

73. Omitted.

77. Omitted.

Exercise 3.3

1. Events E and F **are not** mutually exclusive since a person could be a woman and a doctor.

5. Events E and F **are not** mutually exclusive since a person could have both brown hair and gray hair mixed together or a person could have gray hair dyed brown.

9. Events E and F **are** mutually exclusive since a "four" is not an "odd number."

13. A = card is a ten, B = card is a spade
n(S) = 52, n(A) = 4, n(B) = 13, n(A ∩ B) = 1

    a)   a ten and a spade (A ∩ B):

$$p(A \cap B) = \frac{n(A \cap B)}{n(S)} = \frac{1}{52}$$

13.   Continued

   b)   a ten or a spade $(A \cup B)$:

   $p(A \cup B) = p(A) + p(B) - p(A \cap B)$

   $= \dfrac{n(A)}{n(S)} + \dfrac{n(B)}{n(S)} - \dfrac{n(A \cap B)}{n(S)} = \dfrac{4}{52} + \dfrac{13}{52} - \dfrac{1}{52} = \dfrac{16}{52} = \dfrac{4}{13}$

   c)   not a ten of spades $((A \cap B)')$:

   $p((A \cap B)') = 1 - p(A \cap B) = \dfrac{52}{52} - \dfrac{1}{52} = \dfrac{51}{52}$

17.   A = card is above a five
     A' = card is five or below = {2, 3, 4, 5}
     B = card is below a ten
     B' = card is ten or above = {10, J, Q, K, A}

   a) $n(A') = (4 \text{ suits})(4 \text{ cards}) = 16$, $n(S) = 52$

   $p(A) = 1 - p(A') = 1 - \dfrac{n(A')}{n(S)} = 1 - \dfrac{16}{52} = \dfrac{36}{52} = \dfrac{9}{13}$

   b) $n(B') = (4 \text{ suits})(5 \text{ cards}) = 20$

   $p(B) = 1 - p(B') = 1 - \dfrac{n(B')}{n(S)} = 1 - \dfrac{20}{52} = \dfrac{32}{52} = \dfrac{8}{13}$

   c) $(A \cap B)$ = card is above a five and below a ten
              = {6, 7, 8, 9}

   $n(A \cap B) = (4 \text{ suits})(4 \text{ cards}) = 16$

   $p(A \cap B) = \dfrac{n(A \cap B)}{n(S)} = \dfrac{16}{52} = \dfrac{4}{13}$

   d) $(A \cup B)$ = card is above a five or below a ten
              = card is any card = S

   $p(A \cup B) = p(A) + p(B) - p(A \cap B)$

   $= \dfrac{9}{13} + \dfrac{8}{13} - \dfrac{4}{13} = \dfrac{13}{13} = 1$

   or, $p(A \cup B) = p(S) = 1$

Exercise 3.3

21. F = card is a face card = {J, Q, K}
    n(S) = 52, n(F) = (4 suits)(3 cards) = 12

    $$p(F') = 1 - p(F) = 1 - \frac{n(F)}{n(S)} = 1 - \frac{12}{52} = \frac{52}{52} - \frac{12}{52} = \frac{40}{52} = \frac{10}{13}$$

25. F = card is a jack or higher = {J, Q, K, A}
    n(S) = 52, n(F) = (4 suits)(4 cards) = 16

    $$p(F') = 1 - p(F) = 1 - \frac{n(F)}{n(S)} = 1 - \frac{16}{52} = \frac{52}{52} - \frac{16}{52} = \frac{36}{52} = \frac{9}{13}$$

29. If $p(E) = \frac{2}{7} = \frac{n(E)}{n(S)}$, then n(E) = 2 and n(S) = 7.

    n(E') = n(S) - n(E) = 7 - 2 = 5

    o(E) = n(E):n(E') = 2:5

    o(E') = n(E'):n(E) = 5:2

33. From Exercise 32, if $p(E) = \frac{a}{b}$, then o(E') = (b - a):a.

    K = card is a king
    n(S) = 52, n(K) = (4 suits)(1 card) = 4

    $$p(K) = \frac{n(K)}{n(S)} = \frac{4}{52} = \frac{1}{13} = \frac{a}{b}$$

    o(K') = (b - a):a = (13 - 1):1 = 12:1

37. From Exercise 32, if $p(E) = \frac{a}{b}$, then o(E') = (b - a):a.

    F = card is a four or below = {2, 3, 4}
    n(S) = 52, n(F) = (4 suits)(3 cards) = 12

    $$p(F) = \frac{n(F)}{n(S)} = \frac{12}{52} = \frac{3}{13} = \frac{a}{b}$$

    o(F') = (b - a):a = (13 - 3):3 = 10:3

41. S = shopper who was polled
    A = shopper made a purchase and was happy
    B = shopper made a purchase and was not happy
    C = shopper did not make a purchase and was happy
    D = shopper did not make a purchase and was not happy

    n(S) = 700, n(A) = 151, n(B) = 133, n(C) = 201, n(D) = 215

    a) $p(A) = \frac{n(A)}{n(S)} = \frac{151}{700}$

41. Continued

b) $A \cup B \cup C$ = shopper made a purchase or was happy

$$p(A \cup B \cup C) = p(A) + p(B) + p(C) = \frac{n(A)}{n(S)} + \frac{n(B)}{n(S)} + \frac{n(C)}{n(S)}$$

$$= \frac{151}{700} + \frac{133}{700} + \frac{201}{700} = \frac{485}{700} = \frac{97}{140}$$

45. S = customer who was polled
E = size of bill is between \$40.00 and \$79.99
$n(S) = 1{,}000$, $n(E) = 183 + 177 = 360$

a) $p(E) = \frac{n(E)}{n(S)} = \frac{360}{1000} = \frac{9}{25}$

b) $p(E') = 1 - p(E) = 1 - \frac{n(E)}{n(S)} = 1 - \frac{9}{25} = \frac{25}{25} - \frac{9}{25} = \frac{16}{25}$

49. D = rolling doubles = $\{(1,1), (2,2), (3,3), (4,4), (5,5), (6,6)\}$
E = rolling a 7 = $\{(1,6), (2,5), (3,4), (4,3), (5,2), (6,1)\}$
F = rolling an 11 = $\{(5,6), (6,5)\}$

$n(S) = 36$, $n(D) = 6$, $n(E) = 6$, $n(F) = 2$

a) rolling a 7 or 11, use Rule 5 (E and F are mutually exclusive):

$$p(E \cup F) = p(E) + p(F) = \frac{n(E)}{n(S)} + \frac{n(F)}{n(S)} = \frac{6}{36} + \frac{2}{36} = \frac{8}{36} = \frac{2}{9}$$

b) rolling a 7 or 11 or doubles, use Rule 5 (D, E and F are mutually exclusive):

$$p(D \cup E \cup F) = p(D) + p(E) + p(F) = \frac{n(D)}{n(S)} + \frac{n(E)}{n(S)} + \frac{n(F)}{n(S)}$$

$$= \frac{6}{36} + \frac{6}{36} + \frac{2}{36} = \frac{14}{36} = \frac{7}{18}$$

Exercise 3.3

53. D = rolling doubles = {(1,1), (2,2), (3,3), (4,4), (5,5), (6,6)}
    E = rolling even sums = {(1,1), (1,3), (2,2), (3,1), ..., (6,6)}

    $n(S) = 36$, $n(D) = 6$, $n(E) = 18$ (see figure 3.6)

    a) rolling even and doubles, $(E \cap D) = D$ (All doubles have even sums):

    $$p(E \cap D) = \frac{n(E \cap D)}{n(S)} = \frac{n(D)}{n(S)} = \frac{6}{36} = \frac{1}{6}$$

    b) rolling even or doubles, $(E \cup D)$:

    $$p(E \cup D) = p(E) + p(D) - p(E \cap D)$$

    $$= p(E) + p(D) - p(D) = p(E) = \frac{n(E)}{n(S)} = \frac{18}{36} = \frac{1}{2}$$

57. Let T denote the disease-free gene and t denote the recessive Tay-Sachs disease gene. Create the Punnett squares as indicated.
    D = {(t,t)} (child has the disease)
    C = {(t,T), (T,t)} (child is a carrier)
    $D \cup C$ = {(t,t), (t,T). (T,t)} (child either has the disease or is a carrier)

    a) Each parent is a Tay-Sachs carrier.

	T	t	
	T	t	← carrier parent's genes
T	(T,T)	(t,T)	← offspring
t	(T,t)	(t,t)	← offspring

    ↑
    carrier parent's genes

    $n(S) = 4$, $n(D) = 1$, $n(C) = 2$

    $$p((D \cup C)') = 1 - p(D \cup C) = 1 - (p(D) + p(C))$$

    $$= 1 - \left(\frac{1}{4} + \frac{2}{4}\right) = 1 - \frac{3}{4} = \frac{1}{4}$$

57. Continued

b) One parent is a Tay-Sachs carrier and the othe
no Tay-Sachs gene.

	T	t	← carrier par
T	(T,T)	(t,T)	← offspring
T	(T,T)	(t,T)	← offspring

↑
no Tay-Sachs gene

$n(S) = 4$, $n(D) = 0$, $n(C) = 2$

$$p((D \cup C)') = 1 - p(D \cup C) = 1 - (p(D) + p(c)$$
$$= 1 - \left(\frac{0}{4} + \frac{2}{4}\right) = 1 - \frac{2}{4} = \frac{1}{2}$$

c) One parent has Tay-Sachs and the other parent
Sachs gene.

	t	t	← Tay-Sachs p genes
T	(t,T)	(t,T)	← offspring
T	(t,T)	(t,T)	← offspring

↑
no Tay-Sachs gene

$n(S) = 4$, $n(D) = 0$, $n(C) = 4$

$$p((D \cup C)') = 1 - p(D \cup C) = 1 - [p(D) + p(c)$$
$$= 1 - \left(\frac{0}{4} + \frac{4}{4}\right) = 1 - 1 = 0$$

Exercise 3.3

61. S = flashlights in a large shipment
    A = flashlight has a defective bulb
    B = flashlight has a defective battery
    (A ∩ B) = flashlight has both defects

    p(S) = 1, p(A) = 0.15, p(B) = 0.10, p(A ∩ B) = 0.05

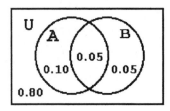

a) p(A ∪ B) = p(A) + p(B) − p(A ∩ B)
    = 0.15 + 0.10 − 0.05 = 0.20

b) p(A' ∪ B') = p((A ∩ B)') = 1 − p(A ∩ B)
    = 1 − 0.05 = 0.95

c) p(A' ∩ B') = p((A ∪ B)') = 1 − p(A ∪ B)
    = 1 − 0.20 = 0.80

65.
$$p(E) + p(E') = \frac{n(E)}{n(S)} + \frac{n(E')}{n(S)} = \frac{n(E)}{n(S)} + \frac{n(S) - n(E)}{n(S)}$$

$$= \frac{n(E) + n(S) - n(E)}{n(S)} = \frac{n(S)}{n(S)} = 1$$

69. a) $\dfrac{18}{33} = \dfrac{3 \cdot 6}{3 \cdot 11} = \dfrac{6}{11}$

    b) Check these results on your calculator.

73. a) $\dfrac{6}{15} + \dfrac{10}{21} = \dfrac{6}{3 \cdot 5} \cdot \left(\dfrac{7}{7}\right) + \dfrac{10}{3 \cdot 7} \cdot \left(\dfrac{5}{5}\right) = \dfrac{42}{105} + \dfrac{50}{105} = \dfrac{92}{105}$

    b) Check these results on your calculator.

77. Omitted.

Exercise 3.4

1. S = all possible lists of 30 birthdays
    E = lists of 30 birthdays in which at least 2 of those birthdays
         are the same
    E' = lists of 30 birthdays in which no two birthdays are the same

    The birthdays may not be repeated so we use **permutations.**

    $_nP_r = \dfrac{n!}{(n - r)!}$, permutation where n = 365, r = 30

    $n(E') = {_{365}P_{30}} = \dfrac{365!}{(365 - 30)!} = \dfrac{365!}{335!} = \dfrac{365 \cdot 364 \cdot 363 \cdot \ldots \cdot 336 \cdot 335!}{335!}$

    $\qquad = 2.1710302 \times 10^{76}$

    $n(S) = 365 \cdot 365 \cdot \ldots \cdot 365 = 365^{30}$

    $p(E) = 1 - p(E') = 1 - \dfrac{n(E')}{n(S)} = 1 - \dfrac{2.1710302 \times 10^{76}}{365^{30}}$

    $\qquad = 0.706316\ldots \approx 0.7063$

5. Winning second prize requires 2 categories:
        5 of the 6 winning numbers (n = 6, r = 5), and
        one that is not a winning number (n = 38, r = 1).

    $_6C_5 = \dfrac{6!}{5!(6 - 5)!} = \dfrac{6!}{5!1!} = \dfrac{6}{1} = 6$

    $_{38}C_1 = \dfrac{38!}{1!(38 - 1)!} = \dfrac{38!}{1!37!} = \dfrac{38}{1} = 38$

    $p\left(\begin{array}{l}\text{second prize in}\\ \text{a 6/44 lottery}\end{array}\right) = \dfrac{n(E)}{n(S)} = \dfrac{_6C_5 \cdot {_{38}C_1}}{_{44}C_6} = \dfrac{6 \cdot 38}{7,059,052}$

    $\qquad\qquad = \dfrac{228}{7,059,052} \approx \dfrac{1}{30,961} \approx 0.00003$

    or, approximately 1 in 31 thousand

Exercise 3.4

9. a) The order is **not** important so use **combinations**

$$_nC_r = \frac{n!}{r!(n-r)!}, \text{ combination where } n = 39, r = 5$$

$$_{39}C_5 = \frac{39!}{5!(39-5)!} = \frac{39!}{5!34!} = \frac{39 \cdot 38 \cdot 37 \cdot 36 \cdot 35}{5 \cdot 4 \cdot 3 \cdot 2 \cdot 1}$$

$$= 575,757$$

E = winning first prize
$n(E) = 1$, $n(S) = {_{39}C_5} = 575,757$

$$p(\text{winning 5/39 lottery}) = \frac{n(E)}{n(S)} = \frac{1}{575,757} \approx 0.000001737$$

or, approximately 1 in 600 thousand.

b) Winning second prize requires 2 categories:
4 of the 5 winning numbers ($n = 5$, $r = 4$), and
one that is not a winning number ($n = 34$, $r = 1$).

$$_5C_4 = \frac{5!}{4!(5-4)!} = \frac{5!}{4!1!} = \frac{5}{1} = 5$$

$$_{34}C_1 = \frac{34!}{1!(34-1)!} = \frac{34!}{1!33!} = \frac{34}{1} = 34$$

$$p\left(\begin{array}{c}\text{second prize in}\\\text{a 5/39 lottery}\end{array}\right) = \frac{n(E)}{n(S)} = \frac{_5C_4 \cdot {_{34}C_1}}{_{39}C_5} = \frac{5 \cdot 34}{575,757}$$

$$= \frac{170}{575,757} \approx \frac{1}{3387} \approx 0.000295$$

or, approximately 1 in 3 thousand

13. The sample space consists of 8 numbers from a possible 80
($n = 80$, $r = 8$). The order is **not** important so use **combinations**.

$$_{80}C_8 = \frac{80!}{8!(80-8)!} = \frac{80!}{8!72!} = 2.899 \times 10^{10}$$

a) Selecting all eight winning spots in eight-spot keno
requires having 8 of a possible 20 winning numbers ($n = 20$,
$r = 8$) combined with none that are not winning ($n = 60$, $r = 0$). The order is **not** important so use **combinations**.

13.a)   Continued

$$_{20}C_8 = \frac{20!}{8!(20-8)!} = \frac{20!}{8!12!} = 125,970$$

$$_{60}C_0 = \frac{60!}{0!(60-0)!} = \frac{60!}{0!60!} = 1$$

$$p(8 \text{ winning spots}) = \frac{_{20}C_8 \cdot {}_{60}C_0}{_{80}C_8} \approx 0.000004$$

b)   Selecting seven winning spots in eight-spot keno requires having 7 of a possible 20 winning numbers (n = 20, r = 7) combined with 1 that is not winning (n = 60, r = 1). The order is **not** important so use **combinations**.

$$_{20}C_7 = \frac{20!}{7!(20-7)!} = \frac{20!}{7!13!} = 77,520$$

$$_{60}C_1 = \frac{60!}{1!(60-1)!} = \frac{60!}{1!59!} = 60$$

$$p(\text{seven winning spots}) = \frac{_{20}C_7 \cdot {}_{60}C_1}{_{80}C_8} \approx 0.000160$$

c)   Selecting six winning spots in eight-spot keno requires having 6 of a possible 20 winning numbers (n = 20, r = 6) combined with 2 that are not winning (n = 60, r = 2). The order is **not** important so use **combinations**.

$$_{20}C_6 = \frac{20!}{6!(20-6)!} = \frac{20!}{6!14!} = 38,760$$

$$_{60}C_2 = \frac{60!}{2!(60-2)!} = \frac{60!}{2!58!} = 1770$$

$$p(\text{six winning spots}) = \frac{_{20}C_6 \cdot {}_{60}C_2}{_{80}C_8} \approx 0.002367$$

13. Continued

    d) Selecting five winning spots in eight-spot keno requires having 5 of a possible 20 winning numbers ($n = 20$, $r = 5$) combined with 3 that are not winning ($n = 60$, $r = 3$). The order is **not** important so use **combinations**.

$$_{20}C_5 = \frac{20!}{5!(20-5)!} = \frac{20!}{5!15!} = 15,504$$

$$_{60}C_3 = \frac{60!}{3!(60-3)!} = \frac{60!}{3!57!} = 34,220$$

$$p(\text{five winning spots}) = \frac{_{20}C_5 \cdot {_{60}C_3}}{_{80}C_8} \approx 0.018303$$

    e) Selecting four winning spots in eight-spot keno requires having 4 of a possible 20 winning numbers ($n = 20$, $r = 4$) combined with 4 that are not winning ($n = 60$, $r = 4$). The order is **not** important so use **combinations**.

$$_{20}C_4 = \frac{20!}{4!(20-4)!} = \frac{20!}{4!16!} = 4845$$

$$_{60}C_4 = \frac{60!}{4!(60-4)!} = \frac{60!}{4!56!} = 487,635$$

$$p(\text{four winning spots}) = \frac{_{20}C_4 \cdot {_{60}C_4}}{_{80}C_8} \approx 0.081504$$

    f) Selecting less than four winning spots in eight-spot keno requires having 1, 2, or 3 of a possible 20 winning numbers. Or, using the Laws of Probabilities, add the above probabilities and subtract from one.

p(less than four winning spots)
    = 1 − [p(8 winning) + p(7 winning) + ... + p(4 winning)]
    = 1 − (0.000004 + 0.000160 + 0.002367 + 0.018303
    + 0.081504)
    = 1 − 0.102338 = 0.897662

17. 12 burritos to go, 5 with hot peppers, 7 without hot peppers. Pick three burritos at random.

   E = all have hot peppers, (n = 5, r = 3)
       implies none without hot peppers (n = 7, r = 0)

   $$_5C_3 = \frac{5!}{3!(5-3)!} = \frac{5!}{3!2!} = \frac{5 \cdot 4}{2 \cdot 1} = 5$$

   $$_7C_2 = \frac{7!}{2!(7-2)!} = \frac{7!}{2!5!} = \frac{7 \cdot 6}{2 \cdot 1} = 21$$

   $$_{12}C_3 = \frac{12!}{3!(12-3)!} = \frac{12!}{3!9!} = \frac{12 \cdot 11 \cdot 10}{3 \cdot 2 \cdot 1} = \frac{1320}{6} = 220$$

   $$p(E) = \frac{_5C_1 \cdot _7C_2}{_{12}C_3} = \frac{(5)(21)}{220} = 0.477273 \approx 0.48$$

21. 12 burritos to go, 5 with hot peppers, 7 without hot peppers. Pick three burritos at random. At most one with hot peppers implies either exactly one with hot peppers or none with hot peppers.

   E = exactly one with hot peppers (n = 5, r = 1)
       implies two without hot peppers (n = 7, r = 2)

   $$_5C_1 = \frac{5!}{1!(5-1)!} = \frac{5!}{1!4!} = \frac{5}{1} = 5$$

   $$_7C_2 = \frac{7!}{2!(7-2)!} = \frac{7!}{2!5!} = \frac{7 \cdot 6}{2 \cdot 1} = 21$$

   $$_{12}C_3 = \frac{12!}{3!(12-3)!} = \frac{12!}{3!9!} = \frac{12 \cdot 11 \cdot 10}{3 \cdot 2 \cdot 1} = \frac{1320}{6} = 220$$

   $$p(E) = \frac{_5C_1 \cdot _7C_2}{_{12}C_3} = \frac{(5)(21)}{220} = \frac{105}{220}$$

Exercise 3.4

21. Continued

F = none with hot peppers (n = 5, r = 0)
implies three without hot peppers (n = 7, r = 3)

$$_5C_0 = \frac{5!}{0!(5-0)!} = \frac{5!}{0!5!} = 1$$

$$_7C_3 = \frac{7!}{3!(7-3)!} = \frac{7!}{3!4!} = \frac{7 \cdot 6 \cdot 5}{3 \cdot 2 \cdot 1} = 35$$

$$_{12}C_3 = \frac{12!}{3!(12-3)!} = \frac{12!}{3!9!} = \frac{12 \cdot 11 \cdot 10}{3 \cdot 2 \cdot 1} = \frac{1320}{6} = 220$$

$$p(E) = \frac{_5C_3 \cdot _7C_0}{_{12}C_3} = \frac{(1)(35)}{220} = \frac{35}{220}$$

$$p\left(\begin{array}{c}\text{at most one}\\\text{hot pepper}\end{array}\right) = p(E) + p(F) = \frac{70}{220} + \frac{10}{220} = \frac{80}{220} \approx 0.36$$

25. The sample space consists of 2 applicants selected from a possible 200. (n = 200, r = 2). The order is **not** important so use **combinations**.

$$_{200}C_2 = \frac{200!}{2!(200-2)!} = \frac{200!}{2!198!} = \frac{200 \cdot 199}{2 \cdot 1} = \frac{39,800}{2} = 19,900$$

a) E = exactly 2 women, (n = 60, r = 2)
implies no men (n = 140, r = 0)

$$_{60}C_2 = \frac{60!}{2!(60-2)!} = \frac{60!}{2!58!} = \frac{60 \cdot 59}{2 \cdot 1} = 1770$$

$$_{140}C_0 = \frac{140!}{0!(140-2)!} = \frac{140!}{0!140!} = \frac{1}{1} = 1$$

$$p(E) = \frac{_{60}C_2 \cdot _{140}C_0}{_{200}C_2} = \frac{(1770)(1)}{19,900} = 0.0889 \approx 0.09$$

b) E = exactly 1 woman, (n = 60, r = 1)
implies one man (n = 140, r = 1)

$$_{60}C_1 = \frac{60!}{1!(60-1)!} = \frac{60!}{1!59!} = \frac{60}{1} = 60$$

$$_{140}C_1 = \frac{140!}{1!(140-1)!} = \frac{140!}{1!139!} = \frac{140}{1} = 140$$

$$p(E) = \frac{_{60}C_1 \cdot _{140}C_1}{_{200}C_2} = \frac{(60)(140)}{19,900} = 0.4221 \approx 0.42$$

25. Continued

   c)   E = exactly 2 men, (n = 140, r = 2)
         implies no women (n = 60, r = 0)

$$_{140}C_2 = \frac{140!}{2!(140-2)!} = \frac{140!}{2!138!} = \frac{140 \cdot 139}{2 \cdot 1} = 9730$$

$$_{60}C_0 = \frac{60!}{0!(60-0)!} = \frac{60!}{0!60!} = \frac{1}{1} = 1$$

$$p(E) = \frac{_{140}C_2 \cdot {}_{60}C_0}{_{200}C_2} = \frac{(9730)(1)}{19,900} = 0.4889 \approx 0.49$$

   d)   Omitted.

Exercise 3.5

1.   E = ball lands on one of two numbers. House odds are 17 to 1.
      (See Figure 3.1.)

$$p(\text{win } E) = \frac{2}{38} = \frac{1}{19}, \ p(\text{lose } E) = 1 - \frac{1}{19} = \frac{18}{19}$$

   expected value = \$17·p(win E) + (-\$1)·p(lose E)

$$= 17 \cdot \frac{1}{19} + (-1) \cdot \frac{18}{19} = \frac{17}{19} + \frac{-18}{19} = \frac{-1}{19} = -\$0.053$$

   On average, \$0.053 would be lost on the two number-bet.

5.   E = ball lands on one of six numbers. House odds are 5 to 1.
      (See Figure 3.1.)

$$p(\text{win } E) = \frac{6}{38} = \frac{3}{19}, \ p(\text{lose } E) = 1 - \frac{3}{19} = \frac{16}{19}$$

   expected value = \$5·p(win E) + (-\$1)·p(lose E)

$$= 5 \cdot \frac{3}{19} + (-1) \cdot \frac{16}{19} = \frac{15}{19} + \frac{-16}{19} = \frac{-1}{19} = -\$0.053$$

   On average, \$0.053 would be lost on the six-number bet.

Exercise 3.5

9. E = ball lands on one of the red numbers. House odds are 1
    to 1. (See Figure 3.1.)

    $p(\text{win E}) = \dfrac{18}{38} = \dfrac{9}{19}$, $p(\text{lose E}) = 1 - \dfrac{9}{19} = \dfrac{10}{19}$

    expected value = \$1•p(win E) + (-\$1)•p(lose E)

    $$= 1•\dfrac{9}{19} + (-1)•\dfrac{10}{19} = \dfrac{9}{19} + \dfrac{-10}{19} = \dfrac{-1}{19} = -\$0.053$$

    On average, \$0.053 would be lost on the red-number bet.

13. Multiply each possible number of books by its probability to
    calculate the expected number of books checked out.

    expected number = 0(0.15) + 1(0.35) + 2(0.25) + 3(0.15)
    $\qquad\qquad\qquad$ + 4(0.05) + 5(0.05)

    $\qquad\qquad$ = 0 + 0.35 + 0.50 + 0.45 + 0.20 + 0.25

    $\qquad\qquad$ = 1.75 books checked out of the library.

17. $35\left(\dfrac{1}{38}\right) + (-1)\left(\dfrac{37}{8}\right) = \dfrac{35 + (-1)(37)}{38} = \dfrac{35 - 37}{38} = \dfrac{-2}{38} \approx -0.053$

21. The pharmaceuticals stock would be the better choice when the
    expected value of the stock is greater than the expected value of
    the bank account which is 0.045. Let p = the probability of
    success for the pharmaceutical stocks, (1 - p) = the probability
    of failure for the stocks.

    expected value of stocks = p(0.5) - (1 - p)(0.6)
    expected value of bank account = 0.045

    expected value of stocks > expected value of bank account

    $\qquad$ (0.5)p - (0.6)(1 - p) > 0.045
    $\qquad\quad$ 0.5p - 0.6 + 0.6p > 0.045
    $\qquad\qquad\quad$ -0.6 + 1.1p > 0.045
    $\qquad\qquad\qquad\quad$ 1.1p > 0.045 + 0.6 = 0.645

    $\qquad\qquad\qquad\qquad$ $p > \dfrac{0.645}{1.1} = 0.586364 \approx 0.59$

    When the probability of success for the pharmaceutical stocks is
    0.59 or more it would be the better choice for Erica.

25. Multiply the probability of each winning bet by the expected profit to calculate the expected value of the bet.

    $$p\text{(four winning spots)} = \frac{{}_{20}C_4 \cdot {}_{60}C_4}{{}_{80}C_8} \approx 0.081504; \quad \text{profits } \$0$$

    $$p\text{(five winning spots)} = \frac{{}_{20}C_5 \cdot {}_{60}C_3}{{}_{80}C_8} \approx 0.018303; \quad \text{profits } \$4$$

    $$p\text{(six winning spots)} = \frac{{}_{20}C_6 \cdot {}_{60}C_2}{{}_{80}C_8} \approx 0.002367; \quad \text{profits } \$99$$

    $$p\text{(seven winning spots)} = \frac{{}_{20}C_7 \cdot {}_{60}C_1}{{}_{80}C_8} \approx 0.000160; \quad \text{profits } \$1479$$

    $$p\text{(8 winning spots)} = \frac{{}_{20}C_8 \cdot {}_{60}C_0}{{}_{80}C_8} \approx 0.000004; \quad \text{profits } \$18,999$$

    $$
    \begin{aligned}
    p\text{(not winning)} &= 1 - p\text{(winning)} \\
    &= 1 - (0.081504 + 0.018303 + \ldots + 0.000004) \\
    &= 1 - 0.102338 = 0.897662; \quad \text{profits } -\$1
    \end{aligned}
    $$

    $$
    \begin{aligned}
    \text{expected value of bet} &= 0(0.081504) + 4(0.018303) + 99(0.002367) \\
    &\quad + 1479(0.000160) + 18,999(0.000004) \\
    &\quad - 1(0.897662) \\
    &= 0.620181 - 0.897662 \\
    &= -\$0.277481 \approx -\$0.28
    \end{aligned}
    $$

29. $p\text{(burning down)} = p = 0.01$
    $p\text{(not burning down)} = 1 - p = 1 - 0.01 = 0.99$

    $$
    \begin{aligned}
    \text{expected value of loss} &= 120,000 \cdot p + (0) \cdot (1 - p) \\
    &= 120,000(0.01) + (0)(0.99) = \$1200
    \end{aligned}
    $$

    The annual premium should be more than $1200.

Exercise 3.5

33. $p(\text{win}) = \dfrac{n(\text{tickets purchased})}{n(\text{possible combinations})} = \dfrac{5,000,000}{7,000,000} = \dfrac{5}{7}$

$p(\text{lose}) = \dfrac{n(\text{tickets not purchased})}{n(\text{possible combinations})} = \dfrac{2,000,000}{7,000,000} = \dfrac{2}{7}$

other = management expenses − funds not spent

$= \dfrac{13,000,000}{2} - 1,500,000 = 6,500,000 - 1,500,000$

$= 5,000,000$

expected value = $(27,000,000)p(\text{win}) - (5,000,000)p(\text{lose}) - \text{other}$

$= 27,000,000\left(\dfrac{5}{7}\right) - 5,000,000\left(\dfrac{2}{7}\right) - 5,000,000$

$= 19,285,714 - 1,428,571 - 5,000,000$

$= \$12,857,143$

37. The sixth bet is the last bet since if the gambler lost there would not be enough money left to double the bet.

Bet Number	Bet	Result	Winnings/ Losses	Total Winnings/ Losses	Cash Left
1	$1	lose	−$1	−$1	$99
2	$2	lose	−$2	−$3	$97
3	$4	lose	−$4	−$7	$93
4	$8	lose	−$8	−$15	$85
5	$16	lose	−$16	−$31	$69
6	$32	lose or win	−$32 or +$32	−$63 or +$1	$37 or $101

The gambler could afford six successive losses before there would not be enough money left to place another bet. If the gambler lost each bet except the last one, the winnings would be $1.

Exercise 3.6

1. a) $p(N) = p(\text{response of "no"}) = \dfrac{n(\text{response of "no"})}{n(\text{total responses})}$

$= \dfrac{n(N)}{n(S)} = \dfrac{140}{600} = 0.233333 \approx 0.23 = 23\%$

This means that of the people responding to the survey 23% felt there was not too much violence on television.

b) $p(W) = p(\text{response from women}) = \dfrac{n(\text{response from women})}{n(\text{total responses})}$

$= \dfrac{n(W)}{n(S)} = \dfrac{320}{600} = 0.533333 \approx 0.53 = 53\%$

This means that 53% of the people responding to the survey were women.

c) $p(N|W) = p(\text{response of "no", given from a woman})$

$= \dfrac{n(N \cap W)}{n(W)} = \dfrac{45}{320} = 0.140625 \approx 0.14 = 14\%$

This means that 14% of the women responding to the survey felt there was not too much violence on television.

d) $p(W|N) = p(\text{response from a women, given it was "no"})$

$= \dfrac{n(W \cap N)}{n(N)} = \dfrac{45}{140} = 0.321429 \approx 0.32 = 32\%$

This means that 32% of the people who responded that there was not too much violence on television were women.

e) $p(N \cap W) = p(\text{response of "no" and from a woman})$

$= \dfrac{n(N \cap W)}{n(S)} = \dfrac{45}{600} = 0.075000 \approx 0.08 = 8\%$

This means that 8% of the total people who responded to the survey were people who said there was not too much violence on television and they were women.

Exercise 3.6

1. Continued

    f) p(W ∩ N) = p(response from a woman and response is "no")

    $$= \frac{n(W \cap N)}{n(S)} = \frac{45}{600} = 0.75000 \approx 0.08 = 8\%$$

    This means that 8% of the total people who responded to the survey were women who said there was not too much violence on television.

5. A = driver had accident
    B = driver was 20-24

    n(B) = 16,900,000, n(A ∩ B) = 5,800,000

    $$p(A|B) = \frac{n(A \cap B)}{n(B)} = \frac{5,800,000}{16,900,000} = 0.343195 \approx 0.34$$

9. A = card dealt first is a diamond
    B = card dealt second is a spade
    n(S) = 52, n(A) = 13, n(B) = 13

    a) $$p(A) = \frac{n(A)}{n(S)} = \frac{13}{52} = \frac{1}{4} = 0.25$$

    b) $$p(B|A) = \frac{n(B)}{n(\text{cards left})} = \frac{13}{51} = 0.255$$

    c) $$p(A \cap B) = p(B|A) \cdot p(A) = \frac{13}{51} \cdot \frac{1}{4} = \frac{13}{204} \approx 0.064$$

    d)

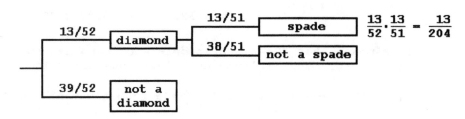

13. a) $p(\text{rolling a 6}) = \dfrac{n(\text{rolling a 6})}{n(S)} = \dfrac{1}{6}$

  b) $p(\text{rolling a 6}|\text{even}) = \dfrac{n(\text{rolling 6 and even})}{n(\text{even})} = \dfrac{1}{3}$

  c) $p(\text{rolling a 6}|\text{odd}) = \dfrac{n(\text{rolling a 6 and odd})}{n(\text{odd})} = \dfrac{0}{3} = 0$

  d) $p(\text{even}|\text{rolling a 6}) = \dfrac{n(\text{even and rolling a 6})}{n(\text{rolling a 6})} = \dfrac{1}{1} = 1$

17. Use Figure 3.6 to calculate the cardinal numbers.

  a) $p(\text{sum} = 4) = \dfrac{n(\text{sum} = 4)}{n(S)} = \dfrac{3}{36} = \dfrac{1}{12}$

  b) $p(\text{sum} = 4|\text{sum} < 6) = \dfrac{n(\text{sum} = 4 \text{ and sum} < 6)}{n(\text{sum} < 6)} = \dfrac{3}{10}$

  c) $p(\text{sum} < 6|\text{sum} = 4) = \dfrac{n(\text{sum} < 6 \text{ and sum} = 4)}{n(\text{sum} = 4)} = \dfrac{3}{3} = 1$

21. S = shopper who was polled
  A = shopper made a purchase and was happy with service
  B = shopper made a purchase and was unhappy with service
  C = shopper made no purchase and was happy with service
  D = shopper made no purchase and was unhappy with service

  $n(S) = 700, \; n(A) = 125, \; n(B) = 111, \; n(C) = 148, \; n(D) = 316$

  Given that a shopper was happy, find the probability of making a purchase.

	Purchase	No Purchase	Total
Happy	$n(A) = 125$	$n(C) = 148$	273

  $p(\text{purchase}|\text{happy}) = \dfrac{n(\text{purchase})}{n(\text{happy})} = \dfrac{125}{273} = 0.4579 \approx 0.46$

25. A = the next four cards after the first card are spades
    B = the first card is a spade

    After each spade is drawn there is one less for the
    numerator and one less for the denominator.

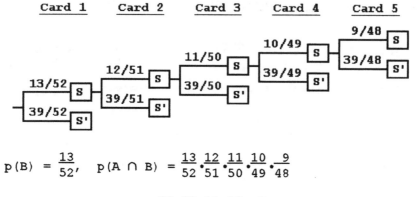

Card 1	Card 2	Card 3	Card 4	Card 5

$$p(B) = \frac{13}{52}, \quad p(A \cap B) = \frac{13}{52} \cdot \frac{12}{51} \cdot \frac{11}{50} \cdot \frac{10}{49} \cdot \frac{9}{48}$$

$$p(A|B) = \frac{p(A \cap B)}{p(B)} = \frac{\frac{13}{52} \cdot \frac{12}{51} \cdot \frac{11}{50} \cdot \frac{10}{49} \cdot \frac{9}{48}}{\frac{13}{52}}$$

$$= \frac{12}{51} \cdot \frac{11}{50} \cdot \frac{10}{49} \cdot \frac{9}{48} = \frac{11,880}{5,997,600} \approx 0.0020$$

29. To calculate probabilities reduce the numerator and denominator
    by one for each branch of the tree.

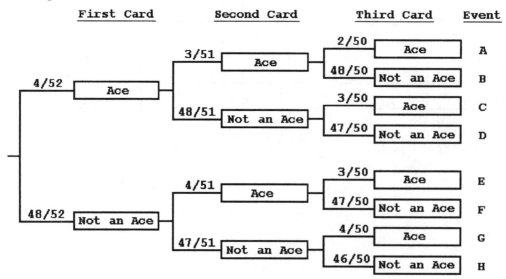

First Card	Second Card	Third Card	Event

29. Continued

p(exactly one ace) = p(D) + p(F) + p(G)

$$= \frac{4}{52} \cdot \frac{48}{51} \cdot \frac{47}{50} + \frac{48}{52} \cdot \frac{4}{51} \cdot \frac{47}{50} + \frac{48}{52} \cdot \frac{47}{51} \cdot \frac{4}{50}$$

$$= 3 \cdot \frac{9,024}{132,600} = 3(0.0680542) \approx 0.20$$

33.

Probabilities Given	Complements of These Probabilities
p(Japan) = 0.38	p(America) = p(not Japan) = 1 − 0.38 = 0.62
p(defective|Japan) = 0.017	p(not defective|Japan) = 1 − 0.017 = 0.983
p(defective|America) = 0.011	p(not defective|America) = 1 − 0.011 = 0.989

p(defective and Japan) = p(defective|Japan)•p(Japan)
= (0.017)(0.38) = .00646 ≈ 0.6%

37. p(pass on first attempt) = 0.61
p(fail on first attempt) = 1 − 0.61 = 0.39
p(pass on second attempt) = 0.63
p(fail on second attempt) = 1 − 0.63 = 0.37
p(pass on third attempt) = 0.42
p(fail on third attempt) = 1 − 0.42 = 0.58

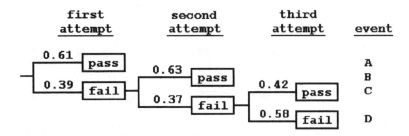

p(a student will pass) = p(A) + p(B) + p(C)
= (0.61) + (0.39)(0.63) + (0.39)(0.37)(0.42)
= 0.61 + 0.2457 + 0.060606
= 0.916306
≈ 92% pass the test

Exercise 3.6

41. A = {ace}
     B = {ten, jack, queen, or king}

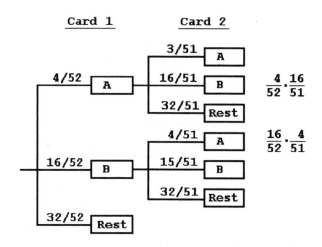

$$p(B|A) = \frac{4}{52} \cdot \frac{16}{51} = \frac{64}{2652}, \quad p(A|B) = \frac{16}{52} \cdot \frac{4}{51} = \frac{64}{2652}$$

$$p(A \cup B) = p(B|A) + p(A|B) = \frac{64}{2652} + \frac{64}{2652}$$

$$= \frac{128}{2652} = 0.048265 \approx 0.05$$

45. a) $p(\text{TB Death}|\text{NYC}) = \dfrac{8,400 + 500}{4,675,000 + 92,000} = \dfrac{8,900}{4,767,000} = 0.00187$

   b) $p(\text{TB Death}|\text{Caucasian NYC}) = \dfrac{8,400}{4,675,000} = 0.00180$

   c) $p(\text{TB Death}|\text{Non-Caucasian NYC}) = \dfrac{500}{92,000} = 0.00543$

   d) $p(\text{TB Death}|\text{Richmond}) = \dfrac{130 + 160}{81,000 + 47,000} = \dfrac{290}{128,000} = 0.00227$

   e) $p(\text{TB Death}|\text{Caucasian Richmond}) = \dfrac{130}{81,000} = 0.00160$

   f) $p(\text{TB Death}|\text{Non-Caucasian Richmond}) = \dfrac{160}{47,000} = 0.00340$

45. Continued

g) Richmond appears to have the higher rate of deaths overall. However, when the data is categorized by race, Richmond has lower rates for both the Caucasian and the non-Caucasian groups. There is a higher rate of TB deaths in the non-Caucasian groups of both cities, but the proportion of non-Caucasians to Caucasians are different in each city causing the total city rates to be reversed. Therefore, New York City has the more severe problem.

	New York City	Richmond
Caucasian	0.00180	0.00160
Non-Caucasian	0.00543	0.00340
Total	0.00187	0.00227

49. $p(A|B) = \dfrac{n(A \cap B)}{n(B)} = \dfrac{\dfrac{n(A \cap B)}{n(S)}}{\dfrac{n(B)}{n(S)}} = \dfrac{p(A \cap B)}{p(B)}$

53. Omitted.

Exercise 3.7

1. a) E and F are **dependent** since knowing that a person is a woman affects the probability that a person is a doctor. There are fewer women doctors than men doctors. $p(E|F) < p(E)$.

   b) E and F are **not mutually exclusive** since you could be both a woman and a doctor simultaneously. $E \cap F \neq \emptyset$.

5. a) Knowing a person has gray hair **does not affect** whether or not that person has brown hair as gray hair could either be dyed brown or some other color or left natural. Therefore, the two events are **independent**. $p(E|F) = p(E)$.

   b) E and F are **not mutually exclusive** since a gray-haired person could dye her hair brown or a person can have both gray and brown hair mixed together. $p(E \cap F) \neq \emptyset$.

Exercise 3.7

9. $p(E) = \frac{1}{6}$, $p(F) = \frac{3}{6}$,

    $p(E \cap F) = 0$ since you cannot get a four and an odd number at the same time.

    $p(E|F) = \dfrac{p(E \cap F)}{p(F)} = \dfrac{0}{\frac{3}{6}} = 0$

    a) E and F are **dependent** since $p(E|F) \neq p(E)$.

    b) E and F are **mutually exclusive** since $E \cap F = \varnothing$.

13. S = shopper who was polled
    A = being happy with the service
    B = making a purchase

	Happy	Not Happy	Total
Purchase	125	111	n(B) = 236
No Purchase	148	316	464
Total	n(A) = 273	427	n(S) = 700

    $p(B) = \dfrac{n(B)}{n(S)} = \dfrac{236}{700} = 0.337$

    $p(B|A) = \dfrac{n(A \cap B)}{n(A)} = \dfrac{125}{273} = 0.458$

    Since $p(B|A) \neq p(B)$, being happy with the service and making a purchase are **dependent** events. We can conclude that people who are happy with the service are more likely to make a purchase.

17. a) p(failure) = 0.01 for a computer **independent** of another computer. Use the product rule for independent events.

    $p(A \cap B \cap C) = p(A) \cdot p(B) \cdot p(C)$

    p(all 3 fail) = p(failure)$\cdot$p(failure)$\cdot$p(failure)
    $$= (0.01)(0.01)(0.01)$$
    $$= 0.000001 \text{ or } 1.0 \times 10^{-6}$$

17. Continued

   b)   3 computers:
$$p(3 \text{ fail}) = (0.01)^3 = 0.000001 = 1.0 \times 10^{-6}$$
(1/1,000,000 or 1 in a million)

   4 computers:
$$p(4 \text{ fail}) = (0.01)^4 = 0.00000001 = 1.0 \times 10^{-8}$$
(1/100,000,000 or 1 in a hundred million)

   5 computers:
$$p(5 \text{ fail}) = (0.01)^5 = 0.0000000001 = 1.0 \times 10^{-10}$$
(1/10,000,000,000 or 1 in 10 billion)

   5 computers would be needed if a 1/1 billion chance of failure is required. Therefore, 4 backup computers are needed.

21. $p(\text{HIV-positive}) = \dfrac{1,000,000}{261,000,000} = 0.003831$

$p(\text{non-HIV-positive}) = 1 - 0.003831 = 0.996169$
$p(\text{SUDS positive}|\text{AIDS/HIV}) = 0.999$
$p(\text{SUDS negative}|\text{non-AIDS/HIV}) = 0.996$

Construct a tree diagram using the above information.

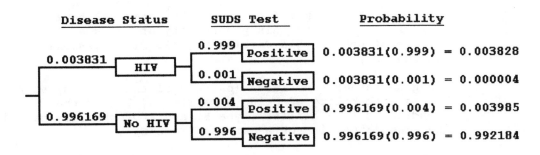

   a) $p(\text{HIV}|+) = \dfrac{0.003828}{0.003828 + 0.003985} = 0.489953 \approx 0.490$

   b) $p(\text{no HIV}|-) = \dfrac{0.992184}{0.992184 + 0.000004} = 0.999996 \approx 1.000$

   c) $p(\text{no HIV}|+) = \dfrac{0.003985}{0.003985 + 0.003828} = 0.510047 \approx 0.510$

21. Continued

    d) $p(HIV|-) = \dfrac{0.000004}{0.000004 + 0.992184} = 0.000004 \approx 0.0$

    e) a and c. (If the test was positive, you would want to know how accurate the test is in predicting the presence of HIV virus.)

    f) The false positive is answer c. The false negative is answer d.

    g) Omitted.

    h) Omitted.

25. a) Three Punnett squares are used.

Type A		
Both Parents are Carriers		
	C	c
C	CC	Cc
c	Cc	cc

Type B		
Only One Parent is a Carrier		
	C	c
C	CC	Cc
C	CC	Cc

Type C		
Neither Parent is a Carrier		
	C	C
C	CC	CC
C	CC	CC

The grandfather comes from Type A but does not have cystic fibrosis. Therefore, his probability of being a carrier is 2/3 and of not being a carrier is 1/3. (cc possibility is eliminated.)

The grandfather's children (parents of the cousins) each have a probability of being a carrier of 1/2 if the grandfather is a carrier (Type B) and a probability of 0 if the grandfather is not a carrier (Type C).

Likewise, the cousins each have a probability of being a carrier of 1/2 if the parent is a carrier (Type B) and a probability of 0 if the parent is not a carrier (Type C).

Construct a tree diagram to find the probability of a cousin being a carrier.

25.a)   Continued

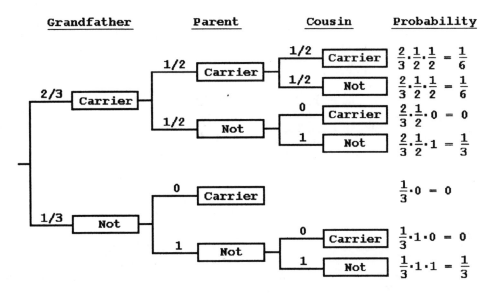

Since the husband (H) and wife (W) are related, the event
that they are both carriers (H ∩ W) is dependent.   There
p(H ∩ W) = p(H|W)·p(W).   From the tree above, the
probability that a cousin, the wife for example, is a
carrier is 1/6.   If the wife is a carrier, the grandfather
must have been a carrier.   The new tree for the husband
follows:

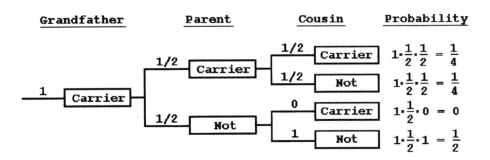

The probability the husband is a carrier given that the wife
is a carrier is 1/4.

$$p(H \cap W) = p(H|W)p(W) = \frac{1}{4} \cdot \frac{1}{6} = \frac{1}{24}$$

Exercise 3.7

25.a)   Continued

Therefore, the probability of both cousins being a carrier would be 1/24. And the probability of their child having cystic fibrosis (Type A) would be 1/96 from the following tree diagram.

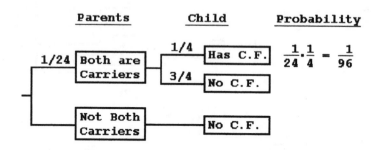

b)   Since husband and wife are unrelated, the probabilities of being carriers are independent. The husband and wife both have the same type of carrier in their background so p(H) = p(W) = 1/6.

$$p(H \cap W) = p(H|W)p(W) = \frac{1}{6} \cdot \frac{1}{6} = \frac{1}{36}$$

And the probability of their child having cystic fibrosis would be 1/144 as seen in the following tree diagram.

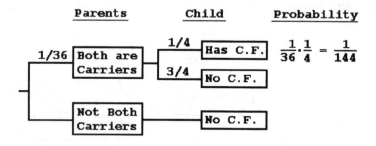

29. Mr. Jones has genes aa. Ms. Jones could have genes Aa or AA with a p(Aa) = 2/3 and p(AA) = 1/3 since we know she is not aa. See Type A in Exercise 25.a).

Possible Punnett squares are:

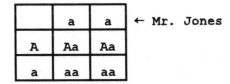

	a	a
A	Aa	Aa
a	aa	aa

← Mr. Jones

↑
Ms. Jones (carrier)

	a	a
A	Aa	Aa
A	Aa	Aa

← Mr. Jones

↑
Ms. Jones (not a carrier)

$$p(Aa) = \frac{2}{4} = \frac{1}{2}$$

$$p(aa) = \frac{2}{4} = \frac{1}{2}$$

$$p(Aa) = \frac{4}{4} = 1$$

$$p(aa) = \frac{0}{4} = 0$$

D = albino child (gene is aa)
E = mother is a carrier (gene is Aa), p(E) = 2/3
F = mother is not a carrier (gene is AA), p(F) = 1/3

$$p(D) = p(D|E) \cdot p(E) + p(D|F) \cdot p(F)$$

$$= \frac{1}{2} \cdot \frac{2}{3} + 0 \cdot \frac{1}{3} = \frac{1}{3} + 0 = \frac{1}{3}$$

The probability of their child being an albino would be 1/3 as can also be seen in the following tree diagram.

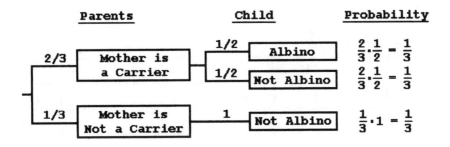

Parents	Child	Probability

33. See Figures 3.24 and 3.25 for gene pairs.

   Mrs. York has black hair so she has genes $M^{Bk}M^{Bk}$ and R^-R^-. Mr. Wilson has dark red hair and genes $M^{Bd}M^{Bw}$ and R^+R^+.

   The Punnett squares are:

Brownness	M^{Bk}	M^{Bk}
M^{Bd}	$M^{Bd}M^{Bk}$	$M^{Bd}M^{Bk}$
M^{Bw}	$M^{Bw}M^{Bk}$	$M^{Bw}M^{Bk}$

Redness	R^-	R^-
R^+	R^+R^-	R^+R^-
R^+	R^+R^-	R^+R^-

$$p(M^{Bd}M^{Bk}) = \frac{2}{4} = \frac{1}{2}, \quad p(M^{Bw}M^{Bk}) = \frac{2}{4} = \frac{1}{2}, \quad p(R^+R^-) = \frac{4}{4} = 1$$

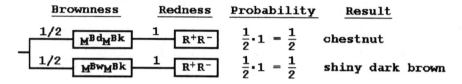

Brownness	Redness	Probability	Result
1/2 $M^{Bd}M^{Bk}$	1 R^+R^-	$\frac{1}{2} \cdot 1 = \frac{1}{2}$	chestnut
1/2 $M^{Bw}M^{Bk}$	1 R^+R^-	$\frac{1}{2} \cdot 1 = \frac{1}{2}$	shiny dark brown

   The child has equal probability of receiving either chestnut hair or shiny dark brown hair.

37. See Figures 3.24 and 3.25 for gene pairs.

   Mrs. Landres has chestnut hair so she has genes $M^{Bd}M^{Bk}$ and R^+R^-. Mr. Landres has shiny dark brown hair and genes $M^{Bw}M^{Bk}$ and R^+R^-.

   The Punnett squares are:

Brownness	M^{Bd}	M^{Bk}
M^{Bw}	$M^{bw}M^{Bd}$	$M^{Bw}M^{Bk}$
M^{Bk}	$M^{Bk}M^{Bd}$	$M^{Bk}M^{Bk}$

Redness	R^+	R^-
R^+	R^+R^+	R^+R^-
R^-	R^-R^+	R^-R^-

37. **Continued**

$$p(M^{Bw}M^{Bd}) = \frac{1}{4}, \quad p(M^{Bw}M^{Bk}) = \frac{1}{4}, \quad p(M^{Bk}M^{Bd}) = \frac{1}{4}, \quad p(M^{Bw}M^{Bk}) = \frac{1}{4}$$

$$p(R^+R^+) = \frac{1}{4}, \quad p(R^+R^-) = \frac{2}{4} = \frac{1}{2}, \quad p(R^-R^-) = \frac{1}{4}$$

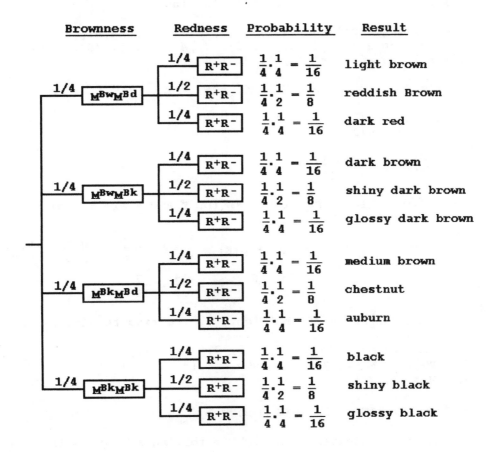

Brownness	Redness	Probability	Result
	1/4 R^+R^-	$\frac{1}{4} \cdot \frac{1}{4} = \frac{1}{16}$	light brown
1/4 $M^{Bw}M^{Bd}$	1/2 R^+R^-	$\frac{1}{4} \cdot \frac{1}{2} = \frac{1}{8}$	reddish Brown
	1/4 R^+R^-	$\frac{1}{4} \cdot \frac{1}{4} = \frac{1}{16}$	dark red
	1/4 R^+R^-	$\frac{1}{4} \cdot \frac{1}{4} = \frac{1}{16}$	dark brown
1/4 $M^{Bw}M^{Bk}$	1/2 R^+R^-	$\frac{1}{4} \cdot \frac{1}{2} = \frac{1}{8}$	shiny dark brown
	1/4 R^+R^-	$\frac{1}{4} \cdot \frac{1}{4} = \frac{1}{16}$	glossy dark brown
	1/4 R^+R^-	$\frac{1}{4} \cdot \frac{1}{4} = \frac{1}{16}$	medium brown
1/4 $M^{Bk}M^{Bd}$	1/2 R^+R^-	$\frac{1}{4} \cdot \frac{1}{2} = \frac{1}{8}$	chestnut
	1/4 R^+R^-	$\frac{1}{4} \cdot \frac{1}{4} = \frac{1}{16}$	auburn
	1/4 R^+R^-	$\frac{1}{4} \cdot \frac{1}{4} = \frac{1}{16}$	black
1/4 $M^{Bk}M^{Bk}$	1/2 R^+R^-	$\frac{1}{4} \cdot \frac{1}{2} = \frac{1}{8}$	shiny black
	1/4 R^+R^-	$\frac{1}{4} \cdot \frac{1}{4} = \frac{1}{16}$	glossy black

The child's possible hair colors and the probability of each color are indicated on the tree diagram above.

41. Omitted.

Chapter 3 Review

1. Omitted.

5. For parts a, b, and c use the following:

    D = rolling doubles = {(1,1), (2,2), (3,3), (4,4), (5,5), (6,6)}
    E = rolling a 7 = {(1,6), (2,5), (3,4), (4,3), (5,2), (6,1)}
    F = rolling an 11 = {(5,6), (6,5)}

    $n(S) = 36$, $n(D) = 6$, $n(E) = 6$, $n(F) = 2$

    a) $p(E) = \dfrac{n(E)}{n(S)} = \dfrac{6}{36} = \dfrac{1}{6}$

    b) $p(F) = \dfrac{2}{36} = \dfrac{1}{18}$

    c)   Using Rule 5 (Section 3.3) (D, E and F are mutually exclusive)

    $$p(D \cup E \cup F) = p(D) + p(E) + p(F) = \dfrac{n(D)}{n(S)} + \dfrac{n(E)}{n(S)} + \dfrac{n(F)}{n(S)}$$

    $$= \dfrac{6}{36} + \dfrac{6}{36} + \dfrac{2}{36} = \dfrac{14}{36} = \dfrac{7}{18}$$

    For parts d, e, and f use Figure 3.6 to find the following:

    G = rolling an odd sum = {3, 5, 7, 9, or 11}
    H = rolling a sum greater than 8 = {9, 10, 11 or 12}
    G ∩ H = {9, 11}

    $n(G) = 18$, $n(H) = 10$, $n(G \cap H) = 6$

    d)   odd and greater than eight = rolling a 9 _or_ an 11

    $$p(G \cap H) = \dfrac{n(G \cap H)}{n(S)} = \dfrac{6}{36} = \dfrac{1}{6}$$

5.   Continued

   e)   odd or greater than eight = rolling a 3, 5, 7, 9 or 11, or a 9, 10, 11 or 12 (not mutually exclusive so use Rule 4 from Section 3.3)

$$p(G \cup H) = p(G) + p(H) - p(G \cap H)$$

$$= \frac{n(G)}{n(S)} + \frac{n(H)}{n(S)} - \frac{n(G \cap H)}{n(S)}$$

$$= \frac{18}{36} + \frac{10}{36} - \frac{6}{36} = \frac{22}{36} = \frac{11}{18}$$

   f)   neither odd nor greater than eight = rolling the complement of part e.

$$p((G \cup H)') = 1 - p(G \cup H)$$

$$= 1 - \frac{22}{36} = \frac{14}{36} = \frac{7}{18}$$

9.   The long-stemmed peas have one gene L and one gene s.   The short-stemmed peas have two genes s.   The Punnett square would look like the following:

	L	s	
			← long-stemmed's genes
s	(s,L)	(s,s)	← offspring
s	(s,L)	(s,s)	← offspring

↑
short-stemmed's genes

   a)   A = {(s,L)}   (long-stemmed offspring)

$$p(A) = \frac{n(A)}{n(S)} = \frac{2}{4} = \frac{1}{2}$$

   b)   B = {(s,s)}   (short-stemmed offspring)

$$p(B) = \frac{n(B)}{n(S)} = \frac{2}{4} = \frac{1}{2}$$

13. One parent is a carrier of Tay-Sachs disease and one parent is not. Let T denote the disease-free gene and t denote the recessive Tay-Sachs disease gene. The Punnett square would look like the following:

	T	t	
			← carrier parent's genes
T	(T,T)	(t,T)	← offspring
T	(T,T)	(t,T)	← offspring

↑
non-carrier parent's genes

a) D = {(t,t)} = ∅ (child has the disease)

$$p(D) = \frac{n(D)}{n(S)} = \frac{0}{4} = 0$$

b) C = {(t,T)} (child is a carrier)

$$p(C) = \frac{n(C)}{n(S)} = \frac{2}{4} = \frac{1}{2}$$

c) H = {(T,T)} (child does not have the disease and is not a carrier)

$$p(H) = \frac{n(H)}{n(S)} = \frac{2}{4} = \frac{1}{2}$$

17. a) Selecting all nine winning spots in nine-spot keno requires having 9 of a possible 20 winning numbers (n = 20, r = 9) combined with none that are not winning (n = 60, r = 0). The sample space requires 9 numbers from a possible 80 (n = 80, r = 9).

The order is **not** important so use **combinations**

$$_{20}C_9 = \frac{20!}{9!(20-9)!} = \frac{20!}{9!11!} = 167,960$$

$$_{60}C_0 = \frac{60!}{0!(60-0)!} = \frac{60!}{0!60!} = 1$$

$$_{80}C_9 = \frac{80!}{9!(80-9)!} = \frac{80!}{9!71!} = 2.319 \times 10^{11}$$

$$p(\text{selecting all 9 winners}) = \frac{_{20}C_9 \cdot _{60}C_0}{_{80}C_9} \approx 0.0000007$$

17. **Continued**

    b) Selecting eight winning spots in nine-spot keno requires
       having 8 of a possible 20 winning numbers ($n = 20$, $r = 8$)
       combined with 1 that is not winning ($n = 60$, $r = 1$). The
       sample space requires 9 numbers from a possible 80 ($n = 80$,
       $r = 9$).

    The order is **not** important so use a **combinations**

    $$_{20}C_8 = \frac{20!}{8!(20-8)!} = \frac{20!}{8!12!} = 125,970$$

    $$_{60}C_1 = \frac{60!}{1!(60-1)!} = \frac{60!}{1!59!} = 60$$

    $$_{80}C_9 = \frac{80!}{9!(80-9)!} = \frac{80!}{9!71!} = 2.319 \times 10^{11}$$

    $$p(\text{selecting eight winners}) = \frac{_{20}C_8 \cdot {_{60}C_1}}{_{80}C_9} \approx 0.00003$$

21. Omitted.

25. Omitted.

4   Statistics

Exercise 4.1

1.   Omitted.

5.   a)   Range = largest - smallest = 80 - 51 = 29

Interval width = $\dfrac{\text{range}}{\text{number of intervals}} = \dfrac{29}{6} = 4.833 \approx 5$

Speed	Tally	Frequency	Relative Frequency
51 ≤ x < 56	||	2	2/40 = 0.05 = 5%
56 ≤ x < 61	||||	4	4/40 = 0.10 = 10%
61 ≤ x < 66	₴₴₴₴ ||	7	7/40 = 0.175 = 17.5%
66 ≤ x < 71	₴₴₴₴ ₴₴₴₴	10	10/40 = 0.25 = 25%
71 ≤ x < 76	₴₴₴₴ ||||	9	9/40 = 0.225 = 22.5%
76 ≤ x < 81	₴₴₴₴ |||	8	8/40 = 0.20 = 20%
		n = 40	total = 100%

b)

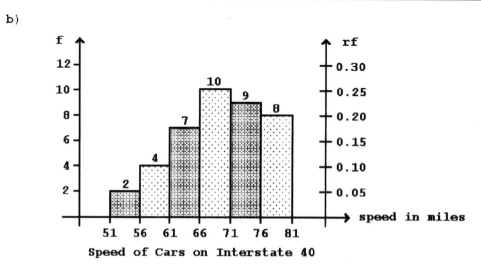

Speed of Cars on Interstate 40

100

9.

Speed	Frequency	Relative Frequency
$4.00 \leq x < 5.50	21	21/152 = 0.138157 = 13.8%
5.50 \leq x < 7.00	35	35/152 = 0.230263 = 23.0%
7.00 \leq x < 8.50	42	42/152 = 0.276316 = 27.6%
8.50 \leq x < 10.00	27	27/152 = 0.177632 = 17.8%
10.00 \leq x < 11.50	18	18/152 = 0.118421 = 11.8%
11.50 \leq x < 13.00	9	9/152 = 0.059211 = 5.9%
	n = 152	total = 99.9% (due to rounding)

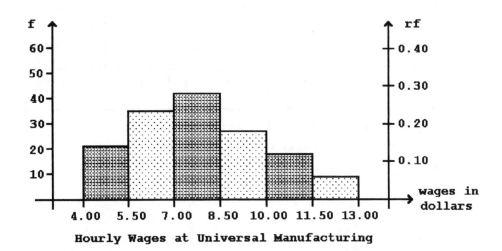

Hourly Wages at Universal Manufacturing

Exercise 4.1

13.

Age of Males	Frequency (in thousands) f	Relative Frequency, f/n	Class Width, w	Relative Frequency Density, rfd = (f/n)÷w
$14 \leq x < 18$	111	111/5893 ≈ 2%	4	(111/5893)÷4 = 0.0047
$18 \leq x < 20$	1380	1380/5893 ≈ 23%	2	(1380/5893)÷2 = 0.1171
$20 \leq x < 22$	1158	1158/5893 ≈ 20%	2	(1158/5893)÷2 = 0.0983
$22 \leq x < 25$	928	928/5893 ≈ 16%	3	(928/5893)÷3 = 0.0525
$25 \leq x < 30$	847	847/5893 ≈ 14%	5	(847/5893)÷5 = 0.0287
$30 \leq x < 35$	591	591/5893 ≈ 10%	5	(591/5893)÷5 = 0.0201
$35 \leq x \leq 60$	878	878/5893 ≈ 15%	25	(878/5893)÷25 = 0.0060
	n = 5893	total = 100%		

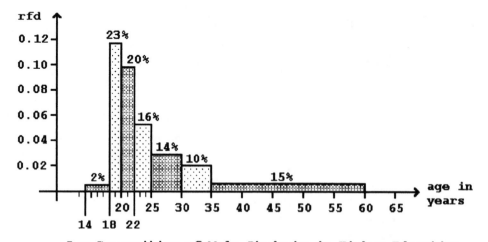

Age Composition of Male Students in Higher Education

102

17.

Grade	Number of students, f	Relative Frequency, f/n	Central Angle (f/n)(360°)
A	6	6/40 = 15%	(6/40)(360°)= 54°
B	16	16/40 = 40%	(16/40)(360°)= 144°
C	12	12/40 = 30%	(12/40)(360°)= 108°
D	4	4/40 = 10%	(4/40)(360°)= 36°
F	2	2/40 = 5%	(2/40)(360°)= 18°
	n = 40	total = 100%	total = 360°

Use a protractor to mark off the appropriate central angles.

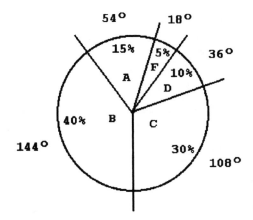

Final Grades in Dr. Gooch's Class

21. Omitted.

Exercise 4.1

25. Enter the data in your calculator. Be sure to change the graphing ranges so that the entire histogram shows on your calculator screen. The histogram should appear similar to this one without the numbers on the x and y axis using the following for the RANGE:

xMin = 0, xMax = 50, xScl = 9, yMin = -2, yMax = 25, yScl = 10

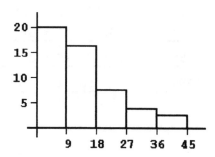

Exercise 4.2

---

1. mean: $\bar{x} = \dfrac{\Sigma x}{n} = \dfrac{9 + 12 + 8 + 10 + \cdots + 10}{14} = \dfrac{175}{14} = 12.5$

    median: Arrange the numbers in order from low to high. Since there are an even number (n = 14) we pick the middle two numbers and find their mean.

    8, 9, 9, 9, 10, 10, 11, 12, 12, 14, 15, 15, 20, 21

    $\dfrac{11 + 12}{2} = 11.5$

    mode: The number with the highest frequency is 9. It occurs three times.

5. a) mean: $\bar{x} = \dfrac{\Sigma x}{n} = \dfrac{9 + 9 + 10 + 11 + 12 + 15}{6} = \dfrac{66}{6} = 11$

    median: Arrange the numbers in order from low to high.
    Since there are an even number (n = 6) we pick the
    middle two numbers and find their mean.

    9, 9, 10, 11, 12, 15
           ↑   ↑
    $\dfrac{10 + 11}{2} = 10.5$

    mode: The number with the highest frequency is 9. It
    occurs two times.

    b) mean: $\bar{x} = \dfrac{\Sigma x}{n} = \dfrac{9 + 9 + 10 + 11 + 12 + 102}{6} = \dfrac{153}{6} = 25.5$

    median: Arrange the numbers in order from low to high.
    Since there are an even number (n = 6) we pick the
    middle two and find their mean.

    9, 9, 10, 11, 12, 102
           ↑   ↑
    $\dfrac{10 + 11}{2} = 10.5$

    mode: The number with the highest frequency is 9. It
    occurs two times.

    c) The means in parts a) and b) are different because the mean
    in part b) has been inflated by the extreme value of 102.
    The medians and the modes in parts a) and b) are equal.

9. mean: $\bar{x} = \dfrac{\Sigma x}{n} = \dfrac{54 + 59 + 35 + \cdots + 22}{15} = \dfrac{659}{15} = 43.933 \approx 43.9$

    median: Arrange the numbers in order from low to high. Since
    there is an odd number (n = 15) we pick the middle
    number.

    22, 25, 34, 35, 41, 41, 46, 46, 46, 47, 49, 54, 54, 59, 60
                                    ↑
    middle value = 46

    mode: 46 has the highest frequency (3) so it is the mode.

Exercise 4.2

13.

x = Weight (in ounces)	Number of Boxes, f	Midpoint, x	f•x
15.3 ≤ x < 15.6	13	15.45	200.85
15.6 ≤ x < 15.9	24	15.75	378.00
15.9 ≤ x < 16.2	84	16.05	1,348.20
16.2 ≤ x < 16.5	19	16.35	310.65
16.5 ≤ x < 16.8	10	16.65	166.50
	n = 150		Σ(f•x) = 2,404.20

mean: $\bar{x} = \dfrac{\Sigma(f•x)}{n} = \dfrac{2,404.2}{150} = 16.028 \approx 16.03$ ounces per box

17. Time to Milwaukee = $\dfrac{90 \text{ mi}}{60 \text{ mph}}$ = 1.5 hr

Time returning = $\dfrac{90 \text{ mi}}{45 \text{ mph}}$ = 2 hr

mean speed = $\dfrac{\text{total distance}}{\text{total time}} = \dfrac{90 + 90}{1.5 + 2} = \dfrac{180}{3.5}$

= 51.428571 ≈ 51.4 mph

21.

Age	Number of People, f	Midpoint, x	f•x
0 < x < 5	18,354,000	2.5	45,855,000
5 ≤ x < 18	45,250,000	11.5	520,375,000
18 ≤ x < 21	11,727,000	19.5	228,676,500
21 ≤ x < 25	15,011,000	23.0	345,253,000
25 ≤ x < 45	80,755,000	35.0	2,826,425,000
45 ≤ x < 55	25,223,000	50.0	1,261,150,000
55 ≤ x < 60	10,532,000	57.5	605,590,000
60 ≤ x < 65	10,616,000	62.5	663,500,000
65 ≤ x < 75	18,107,000	70.0	1,267,490,000
75 ≤ x < 85	10,055,000	80.0	804,400,000
85 ≤ x ≤ 100	3,080,000	92.5	284,900,000
Totals	248,710,000		8,853,644,500

a)   Subtract the "85 ≤ x ≤ 100" row from the totals

$$\bar{x} = \frac{\sum(f \cdot x)}{n} = \frac{8,853,644,500 - 284,900,000}{248,710,000 - 3,080,000}$$

$$= \frac{8,568,744,500}{245,630,000} = 34.884764 \approx 34.9 \text{ mean age}$$

b) $\bar{x} = \dfrac{\sum(f \cdot x)}{n} = \dfrac{8,853,644,500}{248,710,000} = 35.598265 \approx 35.6$ mean age

25. Omitted.

Exercise 4.3

1.

$$n = 6, \; \bar{x} = \frac{\Sigma x}{n} = \frac{42}{6} = 7$$

Data x	Deviation $x - \bar{x}$	Deviation Squared $(x - \bar{x})^2$	x^2
3	3 - 7 = -4	16	9
8	8 - 7 = 1	1	64
5	5 - 7 = -2	4	25
3	3 - 7 = -4	16	9
10	10 - 7 = 3	9	100
13	13 - 7 = 6	36	169
$\Sigma x = 42$		$\Sigma(x - \bar{x})^2 = 82$	$\Sigma x^2 = 376$

a) variance: $s^2 = \dfrac{\Sigma(x - \bar{x})^2}{n - 1} = \dfrac{82}{5} = 16.4$

  standard deviation: $s = \sqrt{\text{variance}} = \sqrt{s^2} = \sqrt{16.4}$

  $$= 4.0497 \approx 4.0$$

b) $s^2 = \dfrac{1}{n - 1}\left(\Sigma x^2 - \dfrac{(\Sigma x)^2}{n}\right)$

  $= \dfrac{1}{6 - 1}\left(376 - \dfrac{(42)^2}{6}\right) = \dfrac{1}{5}(376 - 294) = \dfrac{82}{5} = 16.4$

  $s = \sqrt{\text{variance}} = \sqrt{s^2} = \sqrt{16.4} = 4.0497 \approx 4.0$

5. a)   b)

x	x^2
12	144
16	256
20	400
24	576
28	784
32	1024
$\Sigma x = 132$	$\Sigma x^2 = 3184$

x	x^2
600	360,000
800	640,000
1000	1,000,000
1200	1,440,000
1400	1,960,000
1600	2,560,000
$\Sigma x = 6600$	$\Sigma x^2 = 7,960,000$

a) $n = 6$, $\bar{x} = \dfrac{\Sigma x}{n} = \dfrac{132}{6} = 22$

$$s^2 = \frac{1}{n-1}\left(\Sigma x^2 - \frac{(\Sigma x)^2}{n}\right)$$

$$= \frac{1}{6-1}\left(3184 - \frac{(132)^2}{6}\right) = \frac{1}{5}(3184 - 2904) = \frac{280}{5} = 56$$

$$s = \sqrt{\text{variance}} = \sqrt{s^2} = \sqrt{56} = 7.483315 \approx 7.5$$

b) $n = 6$, $\bar{x} = \dfrac{\Sigma x}{n} = \dfrac{6600}{5} = 1100$

$$s^2 = \frac{1}{n-1}\left(\Sigma x^2 - \frac{(\Sigma x)^2}{n}\right)$$

$$= \frac{1}{6-1}\left(7,960,000 - \frac{(6600)^2}{6}\right)$$

$$= \frac{1}{5}(7,960,000 - 7,260,000) = \frac{700,000}{5} = 140,000$$

$$s = \sqrt{\text{variance}} = \sqrt{s^2} = \sqrt{140,000} = 374.16574 \approx 374.2$$

c) The data in b) represents the data in a) multiplied by 50.

d) The mean in b) is the mean in a) multiplied by 50. The standard deviation in b) is 50 times the standard deviation in a).

Exercise 4.3

9. a)

x = height	x^2
68	4624
50	2500
70	4900
67	4489
72	5184
78	6084
69	4761
68	4624
66	4356
67	4489
$\Sigma x = 675$	$\Sigma x^2 = 46,011$

$n = 10$, $\Sigma x = 675$, $\Sigma x^2 = 46,011$

$$\bar{x} = \frac{\Sigma x}{n} = \frac{675}{10} = 67.5$$

$$s^2 = \frac{1}{n-1}\left(\Sigma x^2 - \frac{(\Sigma x)^2}{n}\right)$$

$$= \frac{1}{10-1}\left(46,011 - \frac{(675)^2}{10}\right) = \frac{1}{9}(46,011 - 45,562.5)$$

$$= \frac{448.5}{9} = 49.833333 \approx 49.8$$

$$s = \sqrt{\text{variance}} = \sqrt{s^2} = \sqrt{49.833333} = 7.0592729 \approx 7.1$$

b) $[\bar{x} - s, \bar{x} + s] = [67.5 - 7.1, 67.5 + 7.1] = [60.4, 74.6]$

Arrange the data in order from smallest to largest to find one standard deviation from the mean.

50, 66, 67, 67, 68, 68, 69, 70, 72, 78

60.4	67.5	74.6
$\bar{x} - s$	$\bar{x}$	$\bar{x} + s$

9.b)   Continued

8 of the 10 observations lie between 60.4 and 74.6.
Therefore, 8/10 = 0.80 or 80% of the data lie within one
standard deviation of the mean.

13.   a)

Score x	Number of Students f	f•x	x^2	f•x^2
10	5	50	100	500
9	10	90	81	810
8	6	48	64	384
7	8	56	49	392
6	3	18	36	108
5	2	10	25	50
Totals	34	272		2244

$n = 34$, $\Sigma(f•x) = 272$, $\Sigma(f•x^2) = 2244$

$$\bar{x} = \frac{\Sigma(f•x)}{n} = \frac{272}{34} = 8$$

$$s^2 = \frac{1}{n-1}\left(\Sigma(f•x^2) - \frac{(\Sigma(f•x))^2}{n}\right)$$

$$= \frac{1}{34-1}\left(2244 - \frac{(272)^2}{34}\right) = \frac{1}{33}(2244 - 2176) = \frac{68}{33}$$

$$= 2.060606 \approx 2.1$$

$$s = \sqrt{\text{variance}} = \sqrt{s^2} = \sqrt{2.060606} = 1.435481 \approx 1.4$$

b) $[\bar{x} - s, \bar{x} + s] = [8 - 1.4, 8 + 1.4] = [6.6, 9.4]$

Scores of 7, 8 and 9 are between 6.6 and 8.4. There are 8 +
6 + 10 = 24 of the 34 students who received that score.
Therefore, 24/34 = 0.70588 or 71% of the data lie within one
standard deviation of the mean.

Exercise 4.3

13. Continued

   c) $[\bar{x} - 2s, \bar{x} + 2s] = [8 - 2(1.4),\ 8 + 2(1.4)]$
       $= [8 - 2.8,\ 8 + 2.8] = [5.2, 10.8]$

   Scores of 6, 7, 8, 9 and 10 are between 5.2 and 10.8. There are $3 + 8 + 6 + 10 + 5 = 32$ of the 34 students who received that score. Therefore, $32/34 = 0.94118$ or 94% of the data lie within two standard deviations of the mean.

   d) $[\bar{x} - 3s, \bar{x} + 3s] = [8 - 3(1.4),\ 8 + 3(1.4)]$
       $= [8 - 4.2,\ 8 + 4.2] = [3.8, 12.2]$

   All the scores are between 3.8 and 12.2. There are 34 of the 34 students who received that score. Therefore, $34/34 = 1.00$ or 100% of the data lie within three standard deviations of the mean.

17. The number of women have been rounded to thousands for the table, but not in the calculations.

Age	Mid-point x	Number of Women f	f•x	$f \cdot x^2 = (f \cdot x)x$
$18 \le x < 25$	21.5	13,167	283,090.5	6,086,445.75
$25 \le x < 30$	27.5	10,839	298,072.5	8,196,993.75
$30 \le x < 35$	32.5	10,838	352,235.0	11,447,637.50
$35 \le x < 40$	37.5	9,586	359,475.0	13,480,312.50
$40 \le x < 45$	42.5	8,155	346,587.5	14,729,968.75
Totals		52,585	1,639,460.5	53,941,358.25

$n = 52,585,000$, $\Sigma(f \cdot x) = 1,639,460,500$,
$\Sigma(f \cdot x^2) = 53,941,358,250$

$$s^2 = \frac{1}{n-1}\left(\Sigma(f \cdot x^2) - \frac{(\Sigma(f \cdot x))^2}{n}\right)$$

$$= \frac{1}{52,585,000 - 1}\left(53,941,358,250 - \frac{(1,639,460,500)^2}{52,585,000}\right)$$

$$= \frac{1}{52,584,999}(53,941,358,250 - 51,114,019,800)$$

17. Continued

$$= \frac{2,827,338,452}{52,584,999} = 53.767015 \approx 53.8$$

$$s = \sqrt{\text{variance}} = \sqrt{s^2} = \sqrt{53.767014} = 7.332599 \approx 7.3$$

21. Using the same data list from Exercise 4.1, problem 25, gives the following data:

$\bar{x} = 13.46$, $\Sigma x = 673$, $\Sigma x^2 = 15473$, $Sx = 11.441439...$,
$\sigma_x = 11.3264...$, $n = 50$

The mean is $\bar{x} = 13.46 \approx 13.5$
The population standard deviation is $\sigma_x = 11.3264... \approx 11.3$

Exercise 4.4

1. a) Use the body table (Appendix VI) to look up 1.00.

    $p(0 < z < 1) = 0.3413$

    So, 34.13% of the z-distribution lies between $z = 0$ and $z = 1$.

    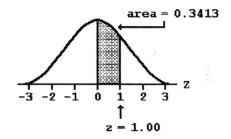

    b) Use the body table to look up 1.00 using the symmetry of the distribution.

    $p(-1 < z < 0)$
    $= p(0 < z < 1) = 0.3413$

    So, 34.13% of the z-distribution lies between $z = -1$ and $z = 0$.

    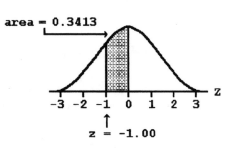

Exercise 4.4

c) Use the body table to look up 1.00, then add the left body to the right body.

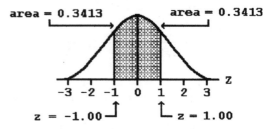

area = 0.3413   area = 0.3413

z = -1.00   z = 1.00

$p(-1 < z < 1) = p(-1 < z < 0) + p(0 < z < 1)$
$= 0.3413 + 0.3413 = 0.6826$

So, 68.26% of the z-distribution lies between z = -1 and z = 1.

5. $\mu = 24.7$, $\sigma = 2.3$

a) One standard deviation of the mean:
$$\mu \pm 1\sigma = 24.7 \pm 1(2.3)$$
$$= 24.7 \pm 2.3$$
$$= [22.4, 27.0]$$
Two standard deviations of the mean:
$$\mu \pm 2\sigma = 24.7 \pm 2(2.3)$$
$$= 24.7 \pm 4.6$$
$$= [20.1, 29.3]$$
Three standard deviations of the mean:
$$\mu \pm 3\sigma = 24.7 \pm 3(2.3)$$
$$= 24.7 \pm 6.9$$
$$= [17.8, 31.6]$$

b) From Figure 4.62
[22.4, 27.0] has 68.26% of the data
[20.1, 29.3] has 95.44% of the data
[17.8, 31.6] has 99.74% of the data

c)

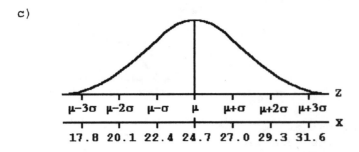

$\mu-3\sigma$ $\mu-2\sigma$ $\mu-\sigma$ μ $\mu+\sigma$ $\mu+2\sigma$ $\mu+3\sigma$ Z

17.8 20.1 22.4 24.7 27.0 29.3 31.6   X

9. a) Find 0.1331 in the interior
        of the body table and read the
        edges to find z-number = 0.34.

        p(0 < z < 0.34) = 0.1331.

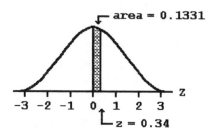

    b) Find 0.4812 in the interior
        of the body table and read
        the edges to find z-number
        = 2.08. Since c is on the
        left of zero, c = -2.08.

        p(-2.08 < z < 0) = 0.4812.

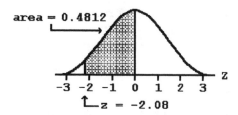

    c) Divide 0.4648 by 2
        and find 0.2324 in
        the interior of the
        body table. Read
        the edges to find
        z-number = 0.62.
        Looking at both
        sides of zero,
        c = 0.62 and
        -c = -0.62.

        p(-0.62 < z < 0.62) = p(-0.62 < z < 0) + p(0 < z < 0.62)
                            = 0.2324 + 0.2324 = 0.4648

    d) p(z > c) = 0.6064
        means c is to the left
        of the mean. Subtract
        0.5 from 0.6064 and
        find 0.1064 in the
        interior of the body
        table. Read the edges
        to find z-number = 0.27
        and change to a
        negative. So, c = -0.27.

        p(z > -0.27) = p(z > 0) + p(-0.27 < z < 0)
                     = 0.5000 + 0.1064 = 0.6064.

Exercise 4.4

9. Continued

e) p(z > c) =
    0.0505, means c
    is to the right
    of the mean.
    Because 50% of
    the distribution
    lies to the right
    of zero, we need
    to subtract

    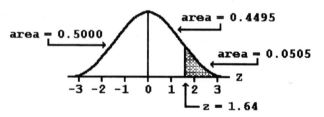

    0.0505 from 0.5 to find the area of the body from z = 0 to
    z = c. Read the edges to find z-number = 1.64. So,
    c = 1.64.

    (z > 1.64) = p(z > 0) − p(0 < z < 1.64)
           = 0.5000 − 0.4495 = 0.0505.

f) p(z < c) = 0.1003
    means c is to the
    left of the mean.
    Because 50% of the
    distribution lies to
    the left of zero, we
    need to subtract
    0.1003 from 0.5 to
    find the area of the

    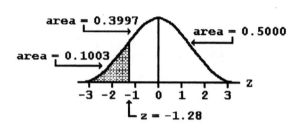

    body from z = 0 to z = c. Read the edges to find z-number =
    1.28 and change it to a negative. So, c = −1.28.

    (z < −1.28) = p(z < 0) − p(−1.28 < z < 0)
           = 0.5000 − 0.3997 = 0.1003.

13. μ = 16.0 ounces, σ = 0.3 ounces

    a) Convert x = 15.5 to a z-number.

    $$z = \frac{x - \mu}{\sigma} = \frac{15.5 - 16.0}{0.3} = \frac{-0.5}{0.3} = -1.67$$

    Find 1.67 on the
    edge of the body
    table and read the
    interior to find
    0.4525. Subtract
    0.4525 from 0.5 to
    find the area to the
    left of z = −1.67.

    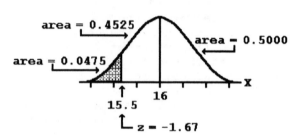

116

13. **Continued**

$$p(x < 15.5) = p(z < -1.67) = p(z < 0) - p(-1.67 < z < 0)$$
$$= 0.5000 - 0.4525 = 0.0475$$

4.75% of the boxes will weigh less than 15.5 ounces.

b) Convert x = 15.8 to a z-number.

$$z = \frac{x - \mu}{\sigma} = \frac{15.8 - 16.0}{0.3} = \frac{-0.2}{0.3} = -0.67$$

Convert x = 16.2 to a z-number.

$$z = \frac{x - \mu}{\sigma} = \frac{16.2 - 16.0}{0.3} = \frac{0.2}{0.3} = 0.67$$

Find 0.67 on the edges of the body table and read the interior to find 0.2486.

Multiply 0.2486 by 2 to get total probability of 0.4972.

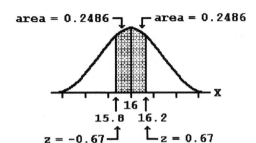

$$p(15.8 < x < 16.2) = p(-0.67 < z < 0.67)$$
$$= 2[p(0 < z < 0.67)] = 2(0.2486) = 0.4972$$

49.72% of the boxes will weigh between 15.8 and 16.2 ounces

17. $\mu = 1.0$ quart, $\sigma = 0.06$ quart

a) Convert x = 0.9 to a z-number.

$$z = \frac{x - \mu}{\sigma} = \frac{0.9 - 1.0}{0.06} = \frac{-0.1}{0.06} = -1.67$$

Find 1.67 on the edges of the body table and read the interior to find 0.4525.

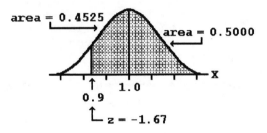

17.a)   Continued

$$p(x > 0.9) = p(z > -1.67) = p(-1.67 < z < 0) + p(z > 0)$$
$$= 0.4525 + 0.5000 = 0.9525$$

95.25% of the cartons will contain at least 0.9 quarts

b)   Convert x = 1.05 to a z-number.

$$z = \frac{x - \mu}{\sigma} = \frac{1.05 - 1.0}{0.06} = \frac{0.05}{0.06} = 0.83$$

Find 0.83 on the edges of
the body table and read
the interior to find
0.2967.

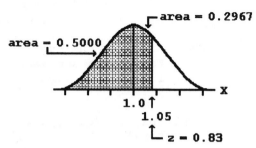

$$p(x < 1.05) = p(z < 0.83) = p(z < 0) + p(0 < z < 0.83)$$
$$= 0.5000 + 0.2967 = 0.7967$$

79.67% of the cartons will contain at most 1.05 quarts

21.   μ = 8.5 minutes, σ = 1.5 minutes

p(z > c) = 0.3400, means c
is to the right of the
mean.   Because 50% of the
distribution lies to the
right of zero, we need to
subtract 0.34 from 0.5 to
find the area of the body
from z = 0 to z = c.   The
closest number to 0.16 is
0.1591 which corresponds to
z-number = 0.41.

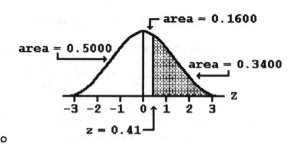

$$p(z > 0.41) = p(z > 0) - p(0 < z < 0.41)$$
$$= 0.5000 - 0.1600 = 0.3400.$$

21. Continued

    Convert z = 0.41 to its x-number.

    $$z = \frac{x - \mu}{\sigma}$$

    $$0.41 = \frac{x - 8.5}{1.5}$$

    $$(1.5)(0.41) = x - 8.5$$
    $$0.615 = x - 8.5$$
    $$x = 9.115 \approx 9.1$$

    $p(z > 0.34) = p(x > 9.1) = 0.3400$

    Extra training will be given if an employee takes 9.1 minutes or longer to package the product.

25. Omitted.

29. Omitted.

Exercise 4.5

1. Find each value in the interior of the body table and read the edges for the z-number.

    a) $z_{0.2517} = 0.68$, that is $p(0 < z < 0.68) = 0.2517$.

    b) $z_{0.1217} = 0.31$, that is $p(0 < z < 0.31) = 0.1217$.

    c) $z_{0.4177} = 1.39$, that is $p(0 < z < 1.39) = 0.4177$.

    d) $z_{0.4960} = 2.65$, that is $p(0 < z < 2.65) = 0.4960$.

5. For a 92% level of confidence, $\alpha = 0.92$, so $\alpha/2 = 0.46$.

    Find 0.46 in the interior of the body table. The closest value is 0.4599 which corresponds to z-number = 1.75.

    $$z_{\alpha/2} = z_{0.46} = 1.75$$

Exercise 4.5

9. n = 1242

    a)   For a 90% level of confidence, α = 0.90, so $\alpha/2$ = 0.45.

         Find 0.45 in the interior of the body table. The closest values are 0.4495 which corresponds to z-number = 1.64 and 0.4505 which corresponds to z-number = 1.65. Use 1.645 which is halfway between the two z-numbers.

         $z_{\alpha/2} = z_{0.45} = 1.645$

         $$MOE = \frac{z_{\alpha/2}}{2\sqrt{n}} = \frac{1.645}{2\sqrt{1242}} = 0.023339 \approx 0.023$$

         We are 90% confident that 25% ($\pm$2.3%) of all Americans do not recall that atomic bombs have ever been used in wartime.

    b)   For a 98% level of confidence, α = 0.98, so $\alpha/2$ = 0.49.

         Find 0.49 in the interior of the body table. The closest value is 0.4901 which corresponds to z-number = 2.33.

         $z_{\alpha/2} = z_{0.49} = 2.33$

         $$MOE = \frac{z_{\alpha/2}}{2\sqrt{n}} = \frac{2.33}{2\sqrt{1242}} = 0.033057 \approx 0.033$$

         We are 98% confident that 25% ($\pm$3.3%) of all Americans do not recall that atomic bombs have ever been used in wartime.

13. n = 2710 = number surveyed
    yes = 2141 = number who have bought a lottery ticket
    no = 569 = number who have not bought a lottery ticket

    a) $p = \frac{yes}{n} = \frac{2141}{2710} = 0.790037 \approx 0.79$

         Sample proportion who bought tickets is 79%.

    b) $p = \frac{no}{n} = \frac{569}{2710} = 0.209963 \approx 0.21$

         Sample proportion who have not purchased tickets is 21%.

13. Continued

   c) For a 90% level of confidence, $\alpha = 0.90$, so $\alpha/2 = 0.45$.

   Find 0.45 in the interior of the body table. The closest values are 0.4495 which corresponds to z-number = 1.64 and 0.4505 which corresponds to z-number = 1.65. Use 1.645 which is halfway between the two z-numbers.

   $$z_{\alpha/2} = z_{0.45} = 1.645$$

   $$\text{MOE} = \frac{z_{\alpha/2}}{2\sqrt{n}} = \frac{1.645}{2\sqrt{2710}} = 0.015800 \approx 0.016$$

   The margin of error associated with the sample proportion is plus or minus 1.6 percentage points at the 90% level of confidence.

17. $n = 2035 = $ teenage Americans surveyed
   yes $= 1160 = $ number who said yes
   no $= 875 = $ number who said no

   a) $$p = \frac{\text{yes}}{n} = \frac{1160}{2035} = 0.570025 \approx 0.57$$

   Sample proportion who know kids who carry weapons in school is 57%.

   b) $$p = \frac{\text{no}}{n} = \frac{875}{2035} = 0.429975 \approx 0.43$$

   Sample proportion who do not know kids who carry weapons in school is 43%.

   c) For a 95% level of confidence, $\alpha = 0.95$, so $\alpha/2 = 0.475$.

   Find 0.475 in the interior of the body table. The closest value is 0.4750 which corresponds to z-number = 1.96.

   $$z_{\alpha/2} = z_{0.475} = 1.96$$

   $$\text{MOE} = \frac{z_{\alpha/2}}{2\sqrt{n}} = \frac{1.96}{2\sqrt{2035}} = 0.021724 \approx 0.022$$

   The margin of error associated with the sample proportion is plus or minus 2.2 percentage points at the 95% level of confidence.

Exercise 4.5

21. For a 95% level of confidence, $\alpha = 0.95$, and $\alpha/2 = 0.475$.

    Find 0.475 in the interior of the body table. The closest value is 0.4750 which corresponds to z-number = 1.96.

    $z_{\alpha/2} = z_{0.475} = 1.96$

    a) n = number of males = 430

    $$MOE = \frac{z_{\alpha/2}}{2\sqrt{n}} = \frac{1.96}{2\sqrt{430}} = 0.047260 \approx 0.047$$

    The margin of error associated with the sample proportion is plus or minus 4.7 percentage points at the 95% level of confidence.

    b) n = number of females = 765

    $$MOE = \frac{z_{\alpha/2}}{2\sqrt{n}} = \frac{1.96}{2\sqrt{765}} = 0.035432 \approx 0.035$$

    The margin of error associated with the sample proportion is plus or minus 3.5 percentage points at the 95% level of confidence.

    c) n = total number = 430 + 765 = 1195

    $$MOE = \frac{z_{\alpha/2}}{2\sqrt{n}} = \frac{1.96}{2\sqrt{1195}} = 0.028349 \approx 0.028$$

    The margin of error associated with the sample proportion is plus or minus 2.8 percentage points at the 95% level of confidence.

25. n = total sample size = 640 + 820 = 1460
    MOE = 2.6% = 0.026

    $$MOE = \frac{z_{\alpha/2}}{2\sqrt{n}}$$

    $$0.026 = \frac{z_{\alpha/2}}{2\sqrt{1460}} = \frac{z_{\alpha/2}}{76.41989}$$

    $z_{\alpha/2} = (0.026)(76.41989) = 1.9869 \approx 1.99$

25. Continued

Find z-number = 1.99 on the edges of the body table and locate
α = 0.4767 in the interior.

If $\alpha/2$ = 0.4767, then α = 2(0.4767) = 0.9534

Based on a sample size of 1460, we are 95.3% confident that the
margin of error is at most 2.6 percentage points.

Exercise 4.6

1. a) n = 6, Σx = 64, Σx^2 = 814, Σy = 85, Σy^2 = 1351, Σxy = 1039

$$m = \frac{n(\Sigma xy) - (\Sigma x)(\Sigma y)}{n(\Sigma x^2) - (\Sigma x)^2} = \frac{6(1039) - (64)(85)}{6(814) - (64)^2}$$

$$= \frac{6234 - 5440}{4884 - 4096} = \frac{794}{788} = 1.00761421$$

$$b = \bar{y} - m\bar{x} = \left(\frac{85}{6}\right) - 1.00761421\left(\frac{64}{6}\right)$$

$$= 14.16666667 - 10.74788494 = 3.41878173$$

$\hat{y}$ = mx + b = 1.00761421x + 3.41878173

Rounding gives $\hat{y}$ = 1.0x + 3.4, the line of best fit.

   b) Substitute x = 11.

$\hat{y}$ = 1.0x + 3.4 = 1.0(11) + 3.4 = 11 + 3.4 = 14.4

   c) Substitute y = 19.

$\hat{y}$ = 1.0x + 3.4

19 = 1.0x + 3.4

x = 19 - 3.4 = 15.6

Exercise 4.6

1.	Continued

	d)	$r = \dfrac{n(\Sigma xy) - (\Sigma x)(\Sigma y)}{\sqrt{n(\Sigma x^2) - (\Sigma x)^2}\ \sqrt{n(\Sigma y^2) - (\Sigma y)^2}}$

	$= \dfrac{6(1039) - (64)(85)}{\sqrt{6(814) - (64)^2}\ \sqrt{6(1351) - (85)^2}}$

	$= \dfrac{794}{\sqrt{4884 - 4096}\ \sqrt{8106 - 7225}} = \dfrac{794}{\sqrt{788}\ \sqrt{881}} = 0.9529485$

	e)	Yes, the predictions in parts b) and c) are reliable because the coefficient of correlation is close to 1.

5.	a)	Yes, the ordered pairs do exhibit a linear trend.

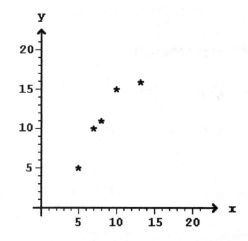

b)

(x, y)	x	x²	y	y²	xy
(5, 5)	5	25	5	25	25
(7, 10)	7	49	10	100	70
(8, 11)	8	64	11	121	88
(10, 15)	10	100	15	225	150
(13, 16)	13	169	16	256	208
n = 5	$\Sigma x = 43$	$\Sigma x^2 = 407$	$\Sigma y = 57$	$\Sigma y^2 = 727$	$\Sigma xy = 541$

5.b) Continued

$$m = \frac{n(\Sigma xy) - (\Sigma x)(\Sigma y)}{n(\Sigma x^2) - (\Sigma x)^2} = \frac{5(541) - (43)(57)}{5(407) - (43)^2}$$

$$= \frac{2705 - 2451}{2035 - 1849} = \frac{254}{186} = 1.36559140$$

$$b = \bar{y} - m\bar{x} = \left(\frac{57}{5}\right) - 1.36559140\left(\frac{43}{5}\right)$$

$$= 11.4 - 11.744408602 = -0.34408602$$

$$\hat{y} = mx + b = 1.36559140x - 0.34408602$$

Rounding gives $\hat{y} = 1.4x - 0.3$, the line of best fit.

c)   Substitute x = 9.

$$\hat{y} = 1.4x - 0.3 = 1.4(9) - 0.3 = 12.6 - 0.3 = 12.3$$

d)

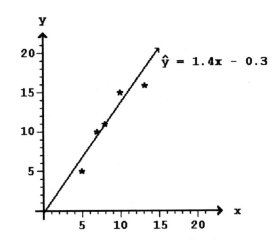

e) $$r = \frac{n(\Sigma xy) - (\Sigma x)(\Sigma y)}{\sqrt{n(\Sigma x^2) - (\Sigma x)^2}\sqrt{n(\Sigma y^2) - (\Sigma y)^2}}$$

$$= \frac{5(541) - (43)(57)}{\sqrt{5(407) - (43)^2}\sqrt{5(727) - (57)^2}}$$

$$= \frac{2705 - 2451}{\sqrt{2035 - 1849}\sqrt{3635 - 3249}} = \frac{254}{\sqrt{186}\sqrt{386}} = 0.9479459$$

Exercise 4.6

5. Continued

   f) Yes, the prediction in part c) is reliable because the coefficient of correlation is close to 1.

9. a) Yes, the data does exhibit a linear trend.

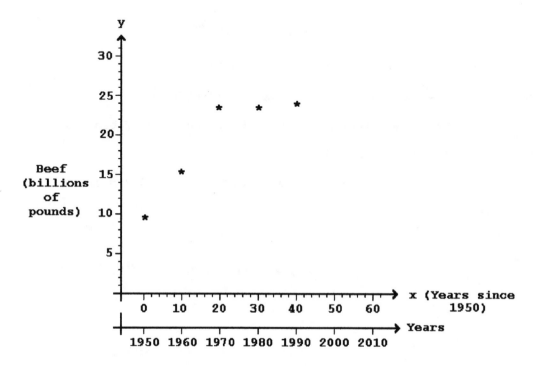

   b)

Year	x	x^2	Beef y	y^2	xy
1950	0	0	9.529	90.801841	0
1960	10	100	15.465	239.166225	154.65
1970	20	400	23.391	547.138881	467.82
1980	30	900	23.513	552.861169	705.39
1990	40	1600	24.031	577.488961	961.24
n = 5	$\Sigma x =$ 100	$\Sigma x^2 =$ 3000	$\Sigma y =$ 95.929	$\Sigma y^2 =$ 2007.457077	$\Sigma xy =$ 2289.10

9.b)   Continued

$$m = \frac{n(\Sigma xy) - (\Sigma x)(\Sigma y)}{n(\Sigma x^2) - (\Sigma x)^2} = \frac{5(2289.10) - (100)(95.929)}{5(3000) - (100)^2}$$

$$= \frac{11,445.5 - 9592.90}{15,000 - 10,000} = \frac{1852.60}{5000} = 0.37052$$

$$b = \bar{y} - m\bar{x} = \left(\frac{95.929}{5}\right) + 0.37052\left(\frac{100}{5}\right)$$

$$= 19.1858 - 7.41040 = 11.77540$$

$$\hat{y} = mx + b = 0.37052x + 11.77540$$

Rounding gives $\hat{y} = 0.3705x + 11.7754$, the line of best fit.

Graphing the line of best fit on the same coordinate system as the data points gives the following:

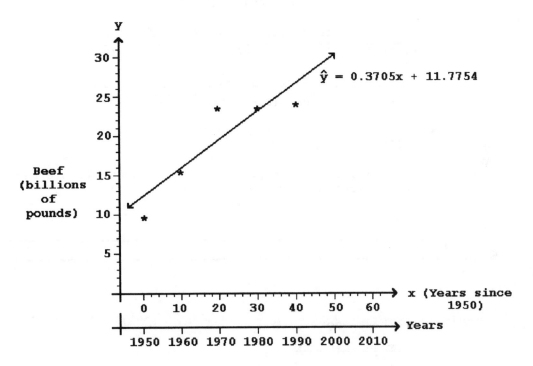

c)   The year 1995 would mean substitute x = 45.

$$\hat{y} = 0.3705(45) + 11.7754 = 16.6725 + 11.7754 = 28.4479$$

Exercise 4.6

9.   Continued

    d)   Substitute y = 30 to predict when 30 billion pounds of beef
          will be consumed.

    $$\hat{y} = 0.3705x + 11.7754$$

    $$30 = 0.3705x + 11.7754$$

    $$0.3705x = 30 - 11.7754 = 18.2246$$

    $$x = \frac{18.2246}{0.3705} = 49.1892038 \approx 49 \text{ years}$$

    Thirty billion pounds of beef will probably be consumed
    in the year 1999 (1950 + 49).

    e) $$r = \frac{n(\Sigma xy) - (\Sigma x)(\Sigma y)}{\sqrt{n(\Sigma x^2) - (\Sigma x)^2} \sqrt{n(\Sigma y^2) - (\Sigma y)^2}}$$

    $$= \frac{5(2289.1) - (100)(95.929)}{\sqrt{5(3000) - (100)^2} \sqrt{5(2007.4571) - (95.929)^2}}$$

    $$= \frac{11,445.5 - 9592.9}{\sqrt{15,000 - 10,000} \sqrt{10,037.28550 - 9202.37304}}$$

    $$= \frac{1852.6}{\sqrt{5000} \sqrt{834.91246}} = 0.9067262$$

    f)   Yes, the predictions in parts c) and d) are reliable because
          the coefficient of correlation is close to 1.

Chapter 4 Review

1.   Omitted.

5.   x = the last test score

    $$\bar{x} = \frac{74 + 65 + 85 + 76 + x}{5} \geq 80$$

    $$\frac{300 + x}{5} \geq 80$$

    $$300 + x \geq (80)(5)$$

5.   Continued

$$300 + x \geq 400$$
$$x \geq 400 - 300 = 100$$

You must score at least 100 on the next exam to average at least 80.

9.   a)   "Weights of motorcycles" are **continuous**, because they would not necessarily be a whole number.

b)   "Colors of motorcycles" are **neither** because they do not have a number value associated with them.

c)   "Number of motorcycles" is **discrete** because you count whole numbers of motorcycles.

d)   "Ethnic background of students" is **neither** because numbers are not associated with backgrounds.

e)   "Number of students" is **discrete** because you count whole numbers of students.

f)   "Amounts of time spent studying" is **continuous** because you can measure time in parts of minutes or seconds.

13. $\mu = 420$, $\sigma = 45$

p(z < c) = 0.3400, means c is to the left of the mean. Because 50% of the distribution lies to the left of zero, we need to subtract 0.34 from 0.5 to find the area of the body from z = 0 to z = c. The closest number to 0.16 is 0.1591 which corresponds to z-number = 0.41. Since c is left of the mean, z-number = -0.41.

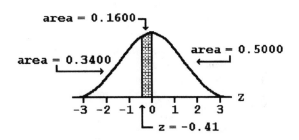

$$p(z < -0.41) = p(z < 0) - p(-0.41 < z < 0)$$
$$= 0.5000 - 0.1600 = 0.3400.$$

Convert z = -0.41 to its x-number.

$$z = \frac{x - \mu}{\sigma}$$

$$-0.41 = \frac{x - 420}{45}$$

13.   Continued

$$(45)(-0.41) = x - 420$$
$$-18.45 = x - 420$$
$$x = 401.55$$

$$p(z < -0.41) = p(x < 401.55) = 0.3400$$

Students who score 401 or lower will have to take the review course. (Do not round to 402 in this situation because a student with that score would not need to take the review course.)

17.   For a 95% level of confidence, $\alpha = 0.95$, and $\alpha/2 = 0.475$.

Find 0.475 in the interior of the body table. The closest value is 0.4750 which corresponds to z-number = 1.96.

$$z_{\alpha/2} = z_{0.475} = 1.96$$

a)   n = number of males = 580

$$MOE = \frac{z_{\alpha/2}}{2\sqrt{n}} = \frac{1.96}{2\sqrt{580}} = 0.040692 \approx 0.041$$

The margin of error associated with the sample proportion is plus or minus 4.1 percentage points at the 95% level of confidence.

b)   n = number of females = 970

$$MOE = \frac{z_{\alpha/2}}{2\sqrt{n}} = \frac{1.96}{2\sqrt{970}} = 0.031466 \approx 0.031$$

The margin of error associated with the sample proportion is plus or minus 3.1 percentage points at the 95% level of confidence.

c)   n = total number = 580 + 970 = 1550

$$MOE = \frac{z_{\alpha/2}}{2\sqrt{n}} = \frac{1.96}{2\sqrt{1150}} = 0.024892 \approx 0.025$$

The margin of error associated with the sample proportion is plus or minus 2.5 percentage points at the 95% level of confidence.

21. a)   Yes, the data does exhibit a linear trend.

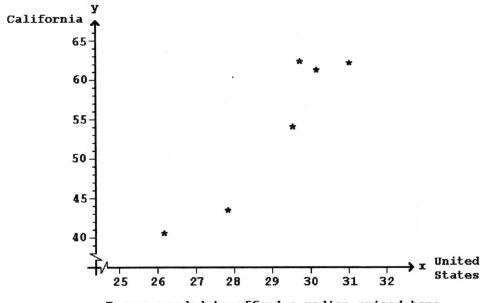

Income needed to afford a median-priced home
(in thousands)

b)   The data is given in thousands of dollars. Data in the
table has been rounded to three decimal places, but all
digits were used in the computations.

Year	U.S. x	x^2	CA y	y^2	xy
1980	26.160	684.346	40.640	1651.610	1063.142
1985	27.842	775.177	43.489	1891.293	1210.821
1989	30.140	908.420	61.330	3761.369	1848.486
1990	29.690	881.496	62.330	3885.029	1850.578
1991	31.011	961.682	62.200	3868.840	1928.884
1992	29.494	869.896	54.058	2922.267	1594.387
n = 6	$\Sigma x =$ 174.337	$\Sigma x^2 =$ 5081.016	$\Sigma y =$ 324.047	$\Sigma y^2 =$ 17,980.408	$\Sigma xy =$ 9496.298

21.b)   Continued

Without rounding the sums are: $n = 6$, $\Sigma x = 174.337$, $\Sigma x^2 = 5081.016421$, $\Sigma y = 324.047$, $\Sigma y^2 = 17,980.40789$, $\Sigma xy = 9496.29789$

$$m = \frac{n(\Sigma xy) - (\Sigma x)(\Sigma y)}{n(\Sigma x^2) - (\Sigma x)^2}$$

$$= \frac{6(9496.29789) - (174.337)(324.047)}{6(5081.016421) - (174.337)^2}$$

$$= \frac{56,977.78734 - 56,493.38184}{30,486.09853 - 30,393.38957} = \frac{484.405501}{92.708957} = 5.2250130$$

$$b = \bar{y} - m\bar{x} = \left(\frac{324.047}{6}\right) + 5.2250130\left(\frac{174.337}{6}\right)$$

$$= 54.00783333 - 151.818847 = -97.8110138, \text{ in thousands}$$

$\hat{y} = mx + b = 5.2250130x - 97,811.0138$

Rounding gives $\hat{y} = 5.225x - 97,811$, the line of best fit.

c)   If the United States income is \$50,000, substitute $x = 50,000$

$\hat{y} = 5.225(50,000) - 97,811$
$= 261,250 - 97,811 = \$163,439$

d)   Substitute $y = 50,000$ to predict the income needed to afford a home in the United States if the income needed in California is \$50,000.

$$\hat{y} = 0.3705x + 11.7754$$

$$50,000 = 5.225x - 97,811$$

$$5.225x = 50,000 + 97,811 = 147,811$$

$$x = \frac{147,811}{5.225} = 28,289.1866 \approx \$28,289$$

21. Continued

e) $r = \dfrac{n(\Sigma xy) - (\Sigma x)(\Sigma y)}{\sqrt{n(\Sigma x^2) - (\Sigma x)^2}\ \sqrt{n(\Sigma y^2) - (\Sigma y)^2}}$

$= \dfrac{6(9496.29789) - (174.337)(324.047)}{\sqrt{6(5081.016) - (174.337)^2}\ \sqrt{6(17,980.408) - (324.047)^2}}$

$= \dfrac{56,977.78734 - 56,493.38184}{\sqrt{30,486.099 - 30,393.390}\ \sqrt{107,882.447 - 1015,006.458}}$

$= \dfrac{484.405501}{\sqrt{92.708957}\ \sqrt{2875.989101}} = 0.93811180$

f)   Yes, the predictions in parts c) and d) are reliable because the coefficient of correlation is close to 1.

5   Finance

Exercise 5.1

It is recommended that the method of using the calculator that is given in the textbook be followed in solving the exercises from this chapter, however, the intermediate calculations have been supplied to assist in checking your work. All decimal digits were used in calculating the final answers even though intermediate steps may be shown as rounded. Final answers and intermediate results may differ slightly depending upon the accuracy of your calculator.

1. $P = \$2000$, $r = 8\% = 0.08$, $t = 3$ years.

   I = Prt   Calculate I
   = (2000)(0.08)(3)   = $480.00

5. $P = \$1410$, $r = 12\frac{1}{4}\% = 0.1225$, $t = 325$ days $= \frac{325}{365}$ years.

   I = Prt   Calculate I
   $= (1410)(0.1225)\left(\frac{325}{365}\right) = 153.79623 \approx \$153.80$

9. $P = \$12,430$, $r = 5\frac{7}{8}\% = 0.05875$,

   $t = 2$ years 3 months $= (24 + 3)$ months $= \frac{27}{12}$ years

   FV = P(1 + rt)   Calculate FV
   $= 12,430\left(1 + (0.05875)\left(\frac{27}{12}\right)\right)$
   $= 12,430(1 + 0.1321875)$
   $= 12,430(1.1321875)$
   $= 14,073.0906 \approx \$14,073.09$

13. $P = \$5900$, $r = 14\frac{1}{2}\% = 0.145$, $t = 112$ days $= \frac{112}{365}$ years

   FV = P(1 + rt)   Calculate FV
   $= 5900\left(1 + (0.145)\left(\frac{112}{365}\right)\right)$
   $= 6162.5096 \approx \$6162.51$

17. $FV = \$8600$, $r = 9\frac{1}{2}\% = 0.095$, $t = 3$ years

$$FV = P(1 + rt) \qquad \text{Calculate P}$$
$$8600 = P(1 + (0.095)(3))$$
$$8600 = P(1.285) \qquad \text{dividing by 1.285 gives}$$
$$P = \frac{8600}{1.285} = 6692.6070 \approx \$6692.61$$

21. $FV = \$1311$, $r = 6\frac{1}{2}\% = 0.065$, $t = 317$ days $= \frac{317}{365}$ years

$$FV = P(1 + rt) \qquad \text{Calculate P}$$
$$1311 = P\left(1 + (0.065)\left(\frac{317}{365}\right)\right)$$
$$1311 = P(1.056452) \qquad \text{dividing by 1.056452 gives}$$
$$P = \frac{1311}{1.056452} = 1240.9460 \approx \$1240.95$$

25. $P = \$1312.82$, $FV = \$1615.00$, $r = 6\frac{7}{8}\% = 0.06875$

$$FV = P(1 + rt) \qquad \text{Calculate t}$$
$$1615.00 = 1312.82(1 + 0.06875t) \qquad \text{divide by 1312.82}$$
$$\frac{1615.00}{1312.82} = 1 + 0.06875t$$
$$1.2301763 = 1 + 0.06875t \qquad \text{subtract 1}$$
$$0.2301763 = 0.06875t \qquad \text{divide by 0.06875}$$
$$\frac{0.2301763}{0.06875} = t$$
$$t = 3.348 \approx 3\frac{1}{3} \text{ years}$$

29. $P = $ Loan Amount $= \$3700 - \$500 = \$3200$, $r = 9.8\% = 0.098$, $t = 36$ months $= 3$ years

$$FV = P(1 + rt) \qquad \text{Calculate FV and then divide by 36}$$
$$FV = 3200(1 + (0.098)(3))$$
$$= \$4140.80$$
$$\text{Monthly Payment} = \frac{4140.80}{36} = 115.0222 \approx \$115.02$$

33. Billing period = March 1 to March 31,
    Previous Balance = $157.14

Time Interval	Days	Daily Balance
March 1 - March 4	4	157.14
March 5 - March 16	12	157.14 - 25.00 = 132.14
March 17 - March 31	15	132.14 + 36.12 = 168.26

$$\text{Average Daily Balance} = \frac{4(157.14) + 12(132.14) + 15(168.26)}{4 + 12 + 15}$$

$$= \frac{4738.14}{31} = 152.8432 \approx \$152.84$$

$P = \$152.8432$, $r = 21\% = 0.21$, $t = 31$ days $= \frac{31}{365}$ years

$I = Prt$ Calculate I

$$= (152.8432)(0.21)\left(\frac{31}{365}\right) = 2.726053 \approx \$2.73$$

37. a) Down Payment = 10% of purchase price
       = 389,400(0.10) = $38,940

    b) Borrowed from bank = 80% of purchase price
       = 389,400(0.80) = $311,520

    c) Borrowed from seller = 10% of purchase price
       = 389,400(0.10) = $38,940

    d) P = 10% of purchase price = 389,400(0.10) = $38,940,

       $r = 11\% = 0.11$, $t = 1$ month $= \frac{1}{12}$ year.

       $I = Prt$ Calculate I

       $$= (38,940)(0.11)\left(\frac{1}{12}\right) = \$356.95$$

    e) Down payment income = 20% of purchase price + monthly
                             interest payments for 48 months

       = (0.20)(389,400) + 48(356.95)

       = 77,880 + 17,133.60 = $95,013.60

37. Continued

  f)   Income from bank = 80% of purchase price – 6% commission

             = (0.80)(389,400) – (0.06)(389,400)

             = 311,520 – 23,364 = $288,156

  g)   Total income = income from bank + down payment income

             = 288,156 + 95,013.60 = $383,169.60

41. Omitted.

45. Omitted.

Exercise 5.2

1. Compound rate = 12% = 0.12

  $$i = \text{periodic rate} = \frac{\text{compound rate}}{\text{number of periods per year}}$$

  a) quarterly = 4 times per year, so $i = \frac{0.12}{4} = 0.03$

  b) monthly = 12 times per year, so $i = \frac{0.12}{12} = 0.01$

  c) daily = 365 times per year, so $i = \frac{0.12}{365} = 0.000328767$

  d) biweekly = 26 times per year, so $i = \frac{0.12}{26} = 0.0046153846$

  e) semimonthly = 24 times per year, so $i = \frac{0.12}{24} = 0.005$

5. Compound Rate = 9.7% = 0.097.   See problem 1 for formula
   and times per year for each section.

  a) quarterly: $i = \frac{0.097}{4} = 0.02425$

  b) monthly: $i = \frac{0.097}{12} = 0.008083333$

  c) daily: $i = \frac{0.097}{365} = 0.000265753$

Exercise 5.2

5.   Continued

d)   biweekly: $i = \dfrac{0.097}{26} = 0.003730769$

e)   semimonthly: $i = \dfrac{0.097}{24} = 0.0040416667$

9.   t = number of years = 30

n = number of periods = (number of years) $\left(\dfrac{\text{number of periods}}{1 \text{ year}}\right)$

a)   quarter: n = (30 years) $\left(\dfrac{4 \text{ quarters}}{1 \text{ year}}\right)$ = 120 quarters

b)   month: n = (30 years) $\left(\dfrac{12 \text{ months}}{1 \text{ year}}\right)$ = 360 months

c)   day: n = (30 years) $\left(\dfrac{365 \text{ days}}{1 \text{ year}}\right)$ = 10,950 days

13.   P = \$5200, $i = \dfrac{6.75\%}{4} = \dfrac{0.0675}{4}$,

n = $\left(8\frac{1}{2} \text{ years}\right)\left(\dfrac{4 \text{ quarters}}{1 \text{ year}}\right)$ = 34 quarters

$\qquad$ FV = P(1 + i)^n $\qquad$ Calculate FV

$\qquad\quad$ = $5200\left(1 + \left(\dfrac{0.0675}{4}\right)\right)^{34}$

$\qquad\qquad$ = 5200(1.016875)^{34}

$\qquad\qquad$ = 5200(1.7664339)

$\qquad\qquad$ = \$9185.4563 $\approx$ \$9185.46

This means the value of \$5200 earning $6\frac{3}{4}\%$ interest compounded quarterly after $8\frac{1}{2}$ years would be \$9185.46.

17. $i = \dfrac{8\%}{12} = \dfrac{0.08}{12}$, $n = \dfrac{12 \text{ months}}{1 \text{ year}} = 12$ months, $t = 1$ year

$$FV\begin{pmatrix} \text{compounded} \\ \text{monthly} \end{pmatrix} = FV\begin{pmatrix} \text{simple} \\ \text{interest} \end{pmatrix}$$

$P(1 + i)^n = P(1 + rt)$ Divide by P, to obtain

$(1 + i)^n = (1 + rt)$ Calculate r

$\left(1 + \dfrac{0.08}{12}\right)^{12} = 1 + r(1)$

$(1.0066667)^{12} = 1 + r$

$1.0829995 = 1 + r$

$r = 0.0829995 \approx 8.30\%$

This means that in 1 year's time 8% interest compounded
monthly would have the same effect as 8.30% simple interest.

21. a) $i = \dfrac{10\%}{4} = \dfrac{0.10}{4}$, $n = \dfrac{4 \text{ quarters}}{1 \text{ year}} = 4$ quarters, $t = 1$ year

$(1 + i)^n = 1 + rt$ Calculate r

$\left(1 + \dfrac{0.10}{4}\right)^4 = 1 + r(1)$

$(1.025)^4 = 1 + r$

$1.1038129 = 1 + r$

$r = 0.1038129 \approx 10.38\%$

This means that in 1 year's time 10% interest compounded
quarterly would have the same effect as 10.38% simple
interest.

b) $i = \dfrac{10\%}{12} = \dfrac{0.10}{12}$, $n = \dfrac{12 \text{ months}}{1 \text{ year}} = 12$ months, $t = 1$ year

$(1 + i)^n = 1 + rt$ Calculate r

$\left(1 + \dfrac{0.10}{12}\right)^{12} = 1 + r(1)$

$(1.00833333)^{12} = 1 + r$

$1.1047131 = 1 + r$

$r = 0.1047131 \approx 10.47\%$

Exercise 5.2

21.b)   Continued

This means that in 1 year's time 10% interest compounded monthly would have the same effect as 10.47% simple interest.

c) $i = \dfrac{10\%}{365} = \dfrac{0.10}{365}$, $n = \dfrac{365 \text{ days}}{1 \text{ year}} = 365$ days, t = 1 year

$$(1 + i)^n = 1 + rt \qquad \text{Calculate } r$$

$$\left(1 + \dfrac{0.10}{365}\right)^{365} = 1 + r(1)$$

$$(1.00027397)^{365} = 1 + r$$

$$1.1051558 = 1 + r$$

$$r = .1051558 \approx 10.52\%$$

This means that in 1 year's time 10% interest compounded daily would have the same effect as 10.52% simple interest.

25.   FV = \$3,758, $i = \dfrac{11\frac{7}{8}\%}{12} = \dfrac{0.11875}{12}$,

n = 17 years$\left(\dfrac{12 \text{ months}}{1 \text{ year}}\right)$ + 7 months = 17(12) + 7 = 211 months

$$FV = P(1 + i)^n \qquad \text{Calculate } P$$

$$3,758 = P\left(1 + \dfrac{0.11875}{12}\right)^{211}$$

$$3,758 = P(1.00989583)^{211}$$

$$3,758 = 7.9865360P$$

$$P = \dfrac{3,758}{7.9865360} = 470.54192 \approx \$470.54$$

This means \$470.54 must be invested now at $11\frac{7}{8}\%$ interest compounded monthly for 17 years 7 months to obtain \$3758.

29. Interest compounded daily: $i = \dfrac{6.5\%}{365} = \dfrac{0.065}{365}$,

n = 365 days, t = 1 year (Calculating annual yield)

FV(compound interest) = FV(simple interest)

$$P(1 + i)^n = P(1 + rt) \qquad \text{Calculate } r$$

$$(1 + i)^n = (1 + rt)$$

$$\left(1 + \frac{0.065}{365}\right)^{365} = 1 + r(1)$$

$$(1.00017808)^{365} = 1 + r$$

$$1.0671529 = 1 + r$$

$$r = 0.0671529 \approx 6.72\% \quad \begin{array}{l}\text{verifying the}\\ \text{advertised yield}\end{array}$$

33. P = \$3000, $r = 6\frac{1}{2}\% = 0.065\%$, $i = \dfrac{0.065}{365}$

a) n = 18 years $\left(\dfrac{365 \text{ days}}{1 \text{ year}}\right)$ = 6570 days

$$FV = P(1 + i)^n \qquad \text{Calculate FV}$$

$$FV = 3000\left(1 + \frac{0.065}{365}\right)^{6570}$$

$$= 3000(1.000178082)^{6570}$$

$$= 9664.9653 \approx \$9664.97$$

b) P = \$9664.97, $i = \dfrac{0.065}{365}$, n = 30 days

$$FV = P(1 + i)^n \qquad \text{Calculate FV}$$

$$FV = 9664.97\left(1 + \frac{0.065}{365}\right)^{30}$$

$$= 9664.97(1.000178082)^{30}$$

$$= 9716.7383 \approx \$9716.74$$

Interest earned = Future Value − Principal
$$= 9716.74 - 9664.97 = \$51.77$$

Exercise 5.2

37. FV = \$100,000 at age 65

a) $i = \dfrac{8\frac{3}{8}\%}{365} = \dfrac{0.08375}{365}$,

$n = (65 - 35 \text{ years})(365 \text{ days}) = 30(365) = 10{,}950 \text{ days}$

$$FV = P(1 + i)^n \qquad \text{Calculate P}$$

$$100{,}000 = P\left(1 + \frac{0.08375}{365}\right)^{10,950}$$

$$100{,}000 = P(1.00022945)^{10,950}$$

$$100{,}000 = P(12.332169)$$

$$P = \frac{100{,}000}{12.332169} = 8108.8738 \approx \$8108.87$$

b) $P = \$100{,}000,\ i = \dfrac{0.08375}{365},\ n = 30 \text{ days}$

$$FV = P(1 + i)^n \qquad \text{Calculate FV}$$

$$FV = 100{,}000\left(1 + \frac{0.08375}{365}\right)^{30}$$

$$= 100{,}000(1.000229452)^{30}$$

$$= 100{,}690.65 \approx \$100{,}690.65$$

$$\text{Interest earned} = \text{Future Value} - \text{Principal}$$

$$= 100{,}690.65 - 100{,}000 = \$690.65$$

41. Find r when t = 1 year.

$$P(1 + i)^n = P(1 + rt)$$

$P(1 + i)^n = P(1 + r(1))$ Substitute t = 1

$(1 + i)^n = 1 + r$ Divide by P

$(1 + i)^n - 1 = r$ Subtract 1

45. $r = 5\frac{5}{8}\% = 0.05625$ Use the formula derived in Exercise 41.

$r = (1 + i)^n - 1$

a) semiannually, $i = \dfrac{0.05625}{2}$, $n = 2$

$r = \left(1 + \dfrac{0.05625}{2}\right)^2 - 1 = (1.028125)^2 - 1$

$= (1.0570410) - 1 = 0.0570410 \approx 5.70\%$

b) quarterly, $i = \dfrac{0.05625}{4}$, $n = 4$

$r = \left(1 + \dfrac{0.05625}{4}\right)^4 - 1 = (1.0140625)^4 - 1$

$= (1.0574477) - 1 = 0.0574477 \approx 5.74\%$

c) monthly, $i = \dfrac{0.05625}{12}$, $n = 12$

$r = \left(1 + \dfrac{0.05625}{12}\right)^{12} - 1 = (1.0046875)^{12} - 1$

$= (1.0577231) - 1 = 0.0577231 \approx 5.77\%$

d) daily, $i = \dfrac{0.05625}{365}$, $n = 365$

$r = \left(1 + \dfrac{0.05625}{365}\right)^{365} - 1 = (1.0001541)^{365} - 1$

$= (1.0578575) - 1 = 0.0578575 \approx 5.79\%$

e) biweekly, $i = \dfrac{0.05625}{26}$, $n = 26$

$r = \left(1 + \dfrac{0.05625}{26}\right)^{26} - 1 = (1.0021635)^{26} - 1$

$= (1.0577978) - 1 = 0.0577978 \approx 5.78\%$

f) semimonthly, $i = \dfrac{0.05625}{2}$, $n = 24$

$r = \left(1 + \dfrac{0.05625}{24}\right)^{24} - 1 = (1.00234375)^{24} - 1$

$= (1.0577925) - 1 = 0.0577925 \approx 5.78\%$

Exercise 5.2

49. Omitted.

53. Omitted.

57. Summary of calculator steps used in this exercise.

Step 1: Determine the two different equations to graph.

$$FV = P(1 + i)^n$$
$$2P = P(1 + i)^n \qquad \text{Future Value} = \text{Double the Principal}$$
$$2 = (1 + i)^n \qquad \text{dividing by P}$$

$$y1 = 2 \quad \text{and} \quad y2 = (1 + i)^x$$

Step 2: Adjust the calculator screen.

Neither x or y can be negative, so xmin = 0, ymin = 0.

Find xmax from doubling time for simple interest.

$$FV = P(1 + rt), \quad r = 0.05$$
$$2P = P(1 + 0.05t)$$
$$2 = 1 + 0.05t \qquad \text{dividing by P}$$
$$1 = 0.05t$$

$$t = \frac{1}{0.05} = 20 \text{ years}$$

xmax = 20 years (number of periods per year)

Find ymax to correspond to xmax.

$$\text{ymax} = (1 + i)^{\text{xmax}}$$

Set xscl and yscl for convenience in reading.

Step 3: Graph. The graphs for this exercise will look similar to the following:

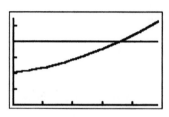

Step 4: Find the intersection. Method will vary depending upon your calculator.

57. Continued

a) compounded annually, $i = \dfrac{0.05}{1} = 0.05$

Step 1: y1 = 2 and y2 = $(1 + 0.05)^x$

Step 2: xmax = 20(1) = 20

ymax = $(1 + 0.05)^{20} = 2.6532 \approx 2.7$

xscl = 2, yscl = 0.5

Step 3: Graph.

Step 4: Intersection: x = 14.206699083, y = 2

Number of periods = 14.2 years

b) compounded monthly, $i = \dfrac{0.05}{12}$

Step 1: y1 = 2 and y2 = $\left(1 + \dfrac{0.05}{12}\right)^x$

Step 2: xmax = 20(12) = 240

ymax = $\left(1 + \dfrac{0.05}{12}\right)^{240} = 2.7126 \approx 2.7$

xscl = 2, yscl = 0.5

Step 3: Graph.

Step 4: Intersection: x = 166.70165675, y = 2

Number of periods = 167 months   or $\dfrac{167}{12} = 13.92$ years

c) compounded quarterly, $i = \dfrac{0.05}{4}$

Step 1: y1 = 2 and y2 = $\left(1 + \dfrac{0.05}{4}\right)^x$

Step 2: xmax = 20(4) = 80

ymax = $\left(1 + \dfrac{0.05}{4}\right)^{80} = 2.701 \approx 2.7$

xscl = 20, yscl = 0.5

Step 3: Graph.

57.c)   Continued

Step 4:   Intersection: $x = 55.797630484$, $y = 2$

Number of periods = 56 quarters   or $\dfrac{56}{4} = 14$ years

d)   compounded daily, $i = \dfrac{0.05}{365}$

Step 1: $y1 = 2$ and $y2 = \left(1 + \dfrac{0.05}{12}\right)^x$

Step 2:   xmax = $20(365) = 7300$

$\text{ymax} = \left(1 + \dfrac{0.05}{365}\right)^{7300} = 2.7181 \approx 2.7$

xscl = 2, yscl = 0.5

Step 3:   Graph.

Step 4:   Intersection: $x = 5060.3209827$, $y = 2$

Number of periods = 5061 days   or $\dfrac{5061}{365} = 13.87$ years

e)   Omitted.

61.   FV = \$30,000, P = \$20,000, $i = \dfrac{6\frac{1}{4}\%}{365} = \dfrac{0.0625}{365}$

$\text{FV} = P(1 + i)^n$ Calculate n

$30,000 = 20,000\left(1 + \dfrac{0.0625}{365}\right)^n$

$1.5 = \left(1 + \dfrac{0.0625}{4}\right)^n$

Equations: $y1 = 1.5$ and $y2 = \left(1 + \dfrac{0.0625}{4}\right)^n$

Simple Interest Approximation:

$30,000 = 20,000(1 + 0.0625t)$
$1.5 = 1 + 0.0625t$
$0.5 = 0.0625t$

$t = \dfrac{0.5}{0.0625} = 8$ years

xmax = $8(365) = 2920 \approx 3000$

61. **Continued**

$$ymax = \left(1 + \frac{0.0625}{4}\right)^{3000} = 1.648 \approx 2.0$$

xscl = 500 and yscl = 0.5

Graph and find the intersection: x = 2368.1189583, y = 1.5

Number of periods = 2369 days or $\frac{2369}{365}$ = 6.49 years

FV = 100,000

$$FV = P(1 + i)^n \qquad \text{Calculate n}$$

$$100,000 = 20,000\left(1 + \frac{0.0625}{365}\right)^n$$

$$5 = \left(1 + \frac{0.0625}{365}\right)^n$$

Equations: $y1 = 5$ and $y2 = \left(1 + \frac{0.0625}{365}\right)^n$

Simple Interest Approximation
$$100,000 = 20,000(1 + 0.0625t)$$
$$5 = 1 + 0.0625t$$
$$4 = 0.0625t$$

$$t = \frac{4}{0.0625} = 64 \text{ years}$$

xmax = 64(365) = 23,360 ≈ 25,000

$$ymax = \left(1 + \frac{0.0625}{4}\right)^{25,000} = 54.5$$

Use ymax = 10, any value over y = 5 would not be critical.
xscl = 5000 and yscl = 1

Graph and find the intersection: x = 9399.9221053, y = 1.5

Number of periods = 9400 days or $\frac{9400}{365}$ = 25.75 years

Exercise 5.3

1. pymt = $120 monthly, $i = \dfrac{5\frac{3}{4}\%}{12} = \dfrac{0.0575}{12}$, n = 1 year = 12 months

$$FV = pymt\frac{(1 + i)^n - 1}{i} \qquad \text{Calculate ordinary FV}$$

$$FV = 120\frac{\left(1 + \dfrac{0.0575}{12}\right)^{12} - 1}{\dfrac{0.0575}{12}}$$

$$= 120\frac{(1.0047917)^{12} - 1}{0.0047917}$$

$$= 120\frac{(1.0590398 - 1)}{0.0047917}$$

$$= 120\left(\frac{0.0590398}{0.0047917}\right)$$

$$= 120(12.321356)$$

$$= 1478.5627 \quad \approx \quad \$1478.56$$

5. pymt = $75 for February to November, $i = \dfrac{7\%}{12} = \dfrac{0.07}{12}$,
    n = 10 months

    a) $FV = pymt\dfrac{(1 + i)^n - 1}{i}$ \qquad Calculate ordinary FV

$$FV = 75\frac{\left(1 + \dfrac{0.07}{12}\right)^{10} - 1}{\dfrac{0.07}{12}}$$

$$= 75(10.266625) = 769.9969 \quad \approx \quad \$770.00$$

    b) total contribution = $75(10 months) = $750.00

    c) total interest = future value - contributions
$$= 770.00 - 750.00 = \$20.00$$

9. pymt = \$175 monthly, $i = \dfrac{10\frac{1}{2}\%}{12} = \dfrac{0.105}{12}$,

   n = (65 − 39 years)(12 months) = 26(12) = 312 months

   a) $FV = \text{pymt}\dfrac{(1 + i)^n - 1}{i}$ Calculate ordinary FV

   $FV = 175\dfrac{\left(1 + \dfrac{0.105}{12}\right)^{312} - 1}{\dfrac{0.105}{12}}$

      = 175(1617.359188) = 283,037.8579 ≈ \$283,037.86

   b) total contribution = \$175(312 months) = \$54,600.00

   c) total interest = future value − contributions

            = 283,037.86 − 54,600.00 = \$228,437.86

13. pymt = \$120 monthly, $i = \dfrac{5\frac{3}{4}\%}{12} = \dfrac{0.0575}{12}$,

   n = 1 year = 12 months,   FV(ordinary) = \$1478.5627

   Future Value (Ordinary) = Future Value (Lump Sum)

              $FV = P(1 + i)^n$ Calculate P

   $1478.5627 = P\left(1 + \dfrac{0.0575}{12}\right)^{12}$

   $1478.5627 = P(1.0047917)^{12}$

   $1478.5627 = P(1.0590398)$

   $P = \dfrac{1478.5627}{1.0590398}$

        = 1396.1351 ≈ \$1396.14

This means that \$1396.14 would need to be deposited at $5\frac{3}{4}\%$ compounded monthly at the beginning of the year to yield the same amount of money (\$1478.56) at the end of the year as with \$120 deposited monthly.

Exercise 5.3

17. pymt = \$175 monthly, $i = \dfrac{10\frac{1}{2}\%}{12} = \dfrac{0.105}{12}$, FV = \$283,037.8579

   $n = (65 - 39 \text{ years})(12 \text{ months}) = 26(12) = 312 \text{ months}$

$$FV = P(1 + i)^n \qquad \text{Calculate P}$$

$$FV = P\left(1 + \frac{0.105}{12}\right)^{312}$$

$$283,037.8579 = P(1.00875)^{312}$$

$$283,037.8579 = P(15.1518929)$$

$$P = \frac{283,037.8579}{15.1518929}$$

$$= 18680.0329 \approx \$18,680.03$$

This means that \$18,680.03 would need to be deposited at $10\frac{1}{2}\%$ compounded monthly at age 39 to yield the same amount of money (\$283,037.86) at age 65 as with \$175 deposited monthly.

21. FV = \$250,000, $i = \dfrac{10\frac{1}{2}\%}{12} = \dfrac{0.105}{12}$,

   $n = (40 \text{ years})(12 \text{ months}) = 40(12) = 480 \text{ months}$

$$FV = \text{pymt}\frac{(1 + i)^n - 1}{i} \qquad \text{Calculate ordinary pymt}$$

$$250,000 = \text{pymt}\frac{\left(1 + \frac{0.105}{12}\right)^{480} - 1}{\frac{0.105}{12}}$$

$$250,000 = \text{pymt}\frac{(1.00875)^{480} - 1}{0.00875}$$

$$250,000 = \text{pymt}\left(\frac{64.4791315}{0.00875}\right)$$

$$250,000 = \text{pymt}(7369.0436)$$

$$\text{pymt} = \frac{250,000}{7369.0436} = 33.925705 \approx \$33.93$$

25. pymt = \$100 biweekly, $i = \dfrac{8\frac{1}{8}\%}{26} = \dfrac{0.08125}{26}$,

$n = 35\frac{1}{2}$ years = (35.5 years)(26 periods) = 923 periods

a) $FV = pymt \dfrac{(1 + i)^n - 1}{i}$ Calculate ordinary FV

$FV = 100\dfrac{\left(1 + \dfrac{0.08125}{26}\right)^{923} - 1}{\dfrac{0.08125}{26}}$

$= 100\dfrac{(1.003125)^{923} - 1}{0.003125}$

$= 100\left(\dfrac{16.8120916}{0.003125}\right)$

$= 100(5379.8693) = 537,986.9319 \approx \$537,986.93$

b) $i = \dfrac{6.1\%}{12} = \dfrac{0.061}{12}$, $n = 1$ month

Interest $= FV - P = P(1 + i)^n - P = P((1 + i)^n - 1)$

$= P\left(\left(1 + \dfrac{0.061}{12}\right)^1 - 1\right)$

$= P(0.005083333)$

<u>Calculations for Month 1 (other months will be similar)</u>

Beginning Account Balance = \$537,986.93

Interest for the Month = P(0.005083333)
                       = 537,986.93(0.005083333)
                       = 2734.7669 $\approx$ \$2734.77

Withdrawal = \$650.00

Ending Account Balance
    = Beginning Balance + Interest − Withdrawal
    = 537,986.93 + 2734.77 − 650.00 = \$540,071.70

25.b) Continued

Month Number	Acct. balance at beginning of month	Interest for the month	Withdrawal	Acct. balance at end of month
1	537,986.93	2734.77	650.00	540,071.70
2	540,071.70	2745.36	650.00	542,167.06
3	542,167.06	2756.02	650.00	544,273.08
4	544,273.08	2766.72	650.00	546,389.80
5	546,389.80	2777.48	650.00	548,517.28

29. $FV = \$1200$, $i = \dfrac{9\%}{12} = \dfrac{0.09}{12}$, $n = (2 \text{ years})(12 \text{ months}) = 24$

$$FV = pymt\,\frac{(1 + i)^n - 1}{i} \qquad \text{Calculate ordinary pymt}$$

$$1200 = pymt\,\frac{\left(1 + \dfrac{0.09}{12}\right)^{24} - 1}{\dfrac{0.09}{12}}$$

$$1200 = pymt\,\frac{(1.0075)^{24} - 1}{0.0075}$$

$$1200 = pymt\left(\frac{0.19641353}{0.0075}\right)$$

$$1200 = pymt\,(26.1884706)$$

$$pymt = \frac{1200}{26.1884706} = 45.8216907 \approx \$45.82$$

33. Future Value (Lump Sum) = Future Value (Ordinary Annuity)

$$P(1 + i)^n = pymt\frac{(1 + i)^n - 1}{i} \qquad \text{Calculate P}$$

Multiply both sides of equation by $\frac{1}{(1 + i)^n} = (1 + i)^{-n}$

$$P(1 + i)^n \left(\frac{1}{(1 + i)^n}\right) = pymt\left(\frac{(1 + i)^n - 1}{i}\right)\left(\frac{1}{(1 + i)^n}\right)$$

giving:

$$P = pymt\frac{(1 + i)^n - 1}{i(1 + i)^n}$$

Factor right side of equation

$$P = pymt\left(\frac{(1 + i)^n}{(1 + i)^n}\right)\left(\frac{1 - (1 + i)^{-n}}{i}\right)$$

and then cancel to give:

$$P = pymt\frac{1 - (1 + i)^{-n}}{i}$$

37. pymt = \$75 for 10 months, $i = \frac{7\%}{12} = \frac{0.07}{12}$, n = 10 months

$$P = pymt\frac{1 - (1 + i)^{-n}}{i} \qquad \text{Calculate P}$$

$$P = 75\frac{1 - \left(1 + \frac{0.07}{12}\right)^{-10}}{\frac{0.07}{12}}$$

$$= 75\frac{1 - (1.0058333)^{-10}}{0.0058333}$$

$$= 75\frac{0.056504659}{0.0058333}$$

$$= 726.4885 \approx \$726.49$$

Exercise 5.3

41. a) FV = \$500,000, pymt = \$200 monthly, $i = \dfrac{5\%}{12} = \dfrac{0.05}{12}$

$$FV = pymt\frac{(1 + i)^n - 1}{i} \qquad \text{Calculate } n$$

$$500,000 = 200\frac{\left(1 + \dfrac{0.05}{12}\right)^n - 1}{\dfrac{0.05}{12}}$$

$$\left(\frac{0.05}{12}\right)(2,500) = \left(1 + \frac{0.05}{12}\right)^n - 1 \qquad \begin{array}{l}\text{divide by 200 and}\\ \text{multiply by denominator}\end{array}$$

$$2,500\left(\frac{0.05}{12}\right) + 1 = \left(1 + \frac{0.05}{12}\right)^n$$

$$y1 = 2,500\left(\frac{0.05}{12}\right) + 1 = 11.4166667, \quad y2 = \left(1 + \frac{0.05}{12}\right)^n$$

Simple interest approximation:
$$y1 = 1 + rt$$
$$11.4166667 = 1 + 0.05t$$
$$10.4166667 = 0.05t$$

$$t = \frac{10.41666667}{0.05} = 208 \text{ years}$$

xmax = 208(12) = 2496
ymax = 15 (any value over 11.4166667 would not be critical)
xscl = 500 and yscl = 5

Graph.
Find the intersection: x = 585.63451963, y = 11.416667

Number of months $= \dfrac{586}{12} = 48$ years 10 months

b) See calculations in a) above.

c)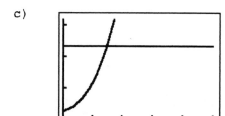

Exercise 5.4

1. $P = \$5000$, $i = \dfrac{9\frac{1}{2}\%}{12} = \dfrac{0.095}{12}$, $n = (4 \text{ years})(12 \text{ months}) = 48 \text{ months}$

   a) <u>monthly payment</u>

$$\text{pymt}\frac{(1 + i)^n - 1}{i} = P(1 + i)^n \qquad \text{Calculate pymt}$$

$$\text{pymt}\frac{\left(1 + \dfrac{0.095}{12}\right)^{48} - 1}{\left(\dfrac{0.095}{12}\right)} = 5000\left(1 + \dfrac{0.095}{12}\right)^{48}$$

$$\text{pymt}\frac{(1.007916667)^{48} - 1}{(0.007916667)} = 5000(1.007916667)^{48}$$

$$\text{pymt}\frac{0.46009824}{0.00791667} = 7300.4912$$

$$\text{pymt}(58.1176726) = 7300.4912$$

$$\text{pymt} = \frac{7300.4912}{58.1176726}$$

$$= 125.6157 \approx \$125.62$$

   b) Total Interest $= \left(\begin{smallmatrix}\text{Number of}\\\text{Payments}\end{smallmatrix}\right)\left(\text{Payment}\right) - \text{Principal}$

   Total Interest $= 48(125.62) - 5000$

$$= 6,029.76 - 5000.00 = \$1,029.76$$

Exercise 5.4

5. $P = \$155,000$, $i = \dfrac{9\frac{1}{2}\%}{12} = \dfrac{0.095}{12}$,

   $n = (30 \text{ years})(12 \text{ months}) = 360 \text{ months}$

   a)   <u>monthly payment</u>

$$\text{pymt}\frac{(1 + i)^n - 1}{i} = P(1 + i)^n \qquad \text{Calculate pymt}$$

$$\text{pymt}\frac{\left(1 + \dfrac{0.095}{12}\right)^{360} - 1}{\left(\dfrac{0.095}{12}\right)} = 155,000\left(1 + \dfrac{0.095}{12}\right)^{360}$$

$$\text{pymt}\frac{(1.007916667)^{360} - 1}{(0.007916667)} = 155,000(1.007916667)^{360}$$

$$\text{pymt}\frac{16.0948614}{.00791667} = 2,649,703.52$$

$$\text{pymt}(2033.03512) = 2,649,703.52$$

$$\text{pymt} = \frac{2,649,703.52}{2033.03512}$$

$$= 1303.3240 \approx \$1303.32$$

   b) $\text{Interest} = \left(\begin{array}{c}\text{Number of}\\\text{Payments}\end{array}\right)\left(\text{Payment}\right) - \text{Principal}$

   $\text{Interest} = 360(1303.32) - 155,000$

   $= 469,195.20 - 155,000.00 = \$314,195.20$

9. $P = \$212,500 - \text{Down Payment} = 212,500 - 212,500(20\%)$

   $= 212,500 - 42,500 = \$170,000$

   $i = \dfrac{10\frac{7}{8}\%}{12} = \dfrac{0.10875}{12}$, $n = 30 \text{ years}(12 \text{ months}) = 360 \text{ months}$

   a)   <u>monthly payment</u>

$$\text{pymt}\frac{(1 + i)^n - 1}{i} = P(1 + i)^n \qquad \text{Calculate pymt}$$

$$\text{pymt}\frac{\left(1 + \dfrac{0.10875}{12}\right)^{360} - 1}{\left(\dfrac{0.10875}{12}\right)} = 170,000\left(1 + \dfrac{0.10875}{12}\right)^{360}$$

9.a) Continued

$$\text{pymt}\frac{(1.0090625)^{360} - 1}{(0.0090625)} = 170,000(1.0090625)^{360}$$

$$\text{pymt}\frac{24.7338053}{0.0090625} = 4,374,746.90$$

$$\text{pymt}(2729.24748) = 4,374,746.90$$

$$\text{pymt} = \frac{4,374,746.90}{2729.24748}$$

$$= 1602.91323 \approx \$1602.91$$

b)   Interest $= \begin{pmatrix}\text{Number of} \\ \text{Payments}\end{pmatrix}\begin{pmatrix}\text{Payment}\end{pmatrix} - \text{Principal}$

Interest $= (360)(1602.91) - 170,000$

$$= 577,047.60 - 170,000.00 = \$407,047.60$$

c)   <u>Amortization Schedule</u>

$r = 10\frac{7}{8}\% = 0.10875$, $t = 1$ month $= \frac{1}{12}$ year

<u>Payment #1</u>

$P = \$170,000.00$

$I = Prt = 170,000(0.10875)\left(\frac{1}{12}\right) = 1540.625 \approx 1540.63$

Principal Portion $=$ Payment $-$ Interest Portion
$$= 1602.91 - 1540.63 = 62.28$$

Amount Due After Payment
$\quad =$ Previous Principal $-$ Principal Portion
$\quad = 170,000 - 62.28 = 169,937.72$

<u>Payment #2</u> (similar to #1)

$P = \$169,937.72$

$I = Prt = 169,937.72(0.10875)\left(\frac{1}{12}\right) = 1540.061 \approx 1540.06$

Principal Portion $=$ Payment $-$ Interest Portion
$$= 1602.91 - 1540.06 = 62.85$$

Amount Due After Payment
$\quad =$ Previous Principal $-$ Principal Portion
$\quad = 169,937.72 - 62.85 = 169,874.87$

Exercise 5.4

9.c) Continued

Payment Number	Principal Portion	Interest Portion	Total Payment	Amount Due After Payment
0	---	---	---	170,000.00
1	62.28	1540.63	1602.91	169,937.72
2	62.85	1540.06	1602.91	169,874.87

d) 38% of monthly income must be greater than monthly payment. Assume that there are only home loan payments.

$$0.38(\text{Income}) > 1602.91$$

$$\text{Income} > \frac{1602.91}{0.38} > 4218.18$$

13. Original Cost = $15,829.32

Loan through the Car Dealer   4 Year Add-On Interest

P = $15,829.32 − Down Payment
  = 15,829.32 − 1000 = 14,829.32

$r = 7\frac{3}{4}\% = 0.0775$, t = 4 years

a) **monthly payment** (Calculate FV and divide by 48)

FV = P(1 + rt)

  = (14,829.32)(1 + 0.0775(4))

  = 14,829.32(1.31) = 19,426.4092

$$\text{Payment} = \frac{19,426.4092}{48} = 404.7169 \approx \$404.72$$

b)   Total Interest = $\left(\begin{matrix}\text{Number of}\\\text{Payments}\end{matrix}\right)\left(\text{Payment}\right)$ − Principal

  = 48(404.72) − 14,829.32

  = 19,426.56 − 14,829.32 = $4597.24

Loan through Bank   4 Year Simple Interest Amortized

P = $15,829.32 − Down Payment = 15,829.32 − 15,829.32(10%)
  = 15,829.32 − 1582.93 = $14,246.39

$i = \dfrac{8\frac{7}{8}\%}{12} = \dfrac{0.08875}{12}$,   n = 4 years(12 months) = 48 months

13. **Continued**

    a) <u>monthly payment</u>

$$\text{pymt}\frac{(1 + i)^n - 1}{i} = P(1 + i)^n \qquad \text{Calculate pymt}$$

$$\text{pymt}\frac{\left(1 + \frac{0.08875}{12}\right)^{48} - 1}{\left(\frac{0.08875}{12}\right)} = (14,246.39)\left(1 + \frac{0.08875}{12}\right)^{48}$$

$$\text{pymt}\frac{(1.00739583)^{48} - 1}{(0.00739583)} = 14,246.39(1.00739583)^{48}$$

$$\text{pymt}\frac{0.42431882}{0.00739583} = 20,291.4013$$

$$\text{pymt}(57.3726851) = 20,291.4013$$

$$\text{pymt} = \frac{20,291.4013}{57.3726851}$$

$$= 353.6770 \approx \$353.68$$

    b) $\text{Interest} = \left(\begin{matrix}\text{Number of} \\ \text{Payments}\end{matrix}\right)\left(\text{Payment}\right) - \text{Principal}$

       $\text{Interest} = (48)(353.68) - 14,246.39$

       $\qquad\qquad = 16,976.64 - 14,246.39 = \$2730.25$

    c) Dennis should choose the simple interest loan from the bank. Although it has a higher down payment, the monthly payments are less and the total amount of interest paid is less.

17. P = $100,000, n = (30 years)(12 months) = 360 months

    a) $i = \frac{6\%}{12} = \frac{0.06}{12}$

$$\text{pymt}\frac{(1 + i)^n - 1}{i} = P(1 + i)^n \qquad \text{Calculate pymt}$$

$$\text{pymt}\frac{\left(1 + \frac{0.06}{12}\right)^{360} - 1}{\left(\frac{0.06}{12}\right)} = 100,000\left(1 + \frac{0.06}{12}\right)^{360}$$

$$\text{pymt}\frac{(1.005)^{360} - 1}{(0.005)} = 100,000(1.005)^{360}$$

17.a) Continued

$$\text{pymt}\frac{5.02257521}{0.005} = 602,257.5212$$

$$\text{pymt}(1004.515042) = 602,25.5212$$

$$\text{pymt} = \frac{602,257.5212}{1004.515042}$$

$$= 599.5505 \approx \$599.55$$

$$\text{Interest} = \binom{\text{Number of}}{\text{Payments}}\Big(\text{Payment}\Big) - \text{Principal}$$

$$\text{Interest} = 360(599.55) - 100,000$$

$$= 215,838.00 - 100,000.00 = \$115,838.00$$

b) $i = \frac{7\%}{12} = \frac{0.07}{12}$

$$\text{pymt}\frac{(1 + i)^n - 1}{i} = P(1 + i)^n \qquad \text{Calculate pymt}$$

$$\text{pymt}\frac{\left(1 + \frac{0.07}{12}\right)^{360} - 1}{\left(\frac{0.07}{12}\right)} = 100,000\left(1 + \frac{0.07}{12}\right)^{360}$$

$$\text{pymt}\frac{(1.00583333)^{360} - 1}{(0.00583333)} = 100,000(1.00583333)^{360}$$

$$\text{pymt}\frac{7.11649747}{0.00583333} = 811,649.747$$

$$\text{pymt}(1219.970996) = 811,649.747$$

$$\text{pymt} = \frac{811,649.747}{1219.970996}$$

$$= 665.3025 \approx \$665.30$$

$$\text{Interest} = \binom{\text{Number of}}{\text{Payments}}\Big(\text{Payment}\Big) - \text{Principal}$$

$$\text{Interest} = 360(665.30) - 100,000$$

$$= 239,508.00 - 100,000.00 = \$139,508.00$$

17. Continued

c) $i = \dfrac{8\%}{12} = \dfrac{0.08}{12}$

$$\text{pymt}\dfrac{(1 + i)^n - 1}{i} = P(1 + i)^n \qquad \text{Calculate pymt}$$

$$\text{pymt}\dfrac{\left(1 + \dfrac{0.08}{12}\right)^{360} - 1}{\left(\dfrac{0.08}{12}\right)} = 100,000\left(1 + \dfrac{0.08}{12}\right)^{360}$$

$$\text{pymt}\dfrac{(1.00666667)^{360} - 1}{(0.00666667)} = 100,000(1.00666667)^{360}$$

$$\text{pymt}\dfrac{9.9357294}{0.00666667} = 1,093,572.94$$

$$\text{pymt}(1,490.35941) = 1,093,572.94$$

$$\text{pymt} = \dfrac{1,093,572.94}{1490.35941}$$

$$= 733.7646 \approx \$733.76$$

$$\text{Interest} = \left(\begin{array}{c}\textbf{Number of}\\\textbf{Payments}\end{array}\right)\left(\textbf{Payment}\right) - \textbf{Principal}$$

$$\text{Interest} = 360(733.76) - 100,000$$

$$= 264,153.60 - 100,000.00 = \$164,153.60$$

d) $i = \dfrac{9\%}{12} = \dfrac{0.09}{12}$

$$\text{pymt}\dfrac{(1 + i)^n - 1}{i} = P(1 + i)^n \qquad \text{Calculate pymt}$$

$$\text{pymt}\dfrac{\left(1 + \dfrac{0.09}{12}\right)^{360} - 1}{\left(\dfrac{0.09}{12}\right)} = 100,000\left(1 + \dfrac{0.09}{12}\right)^{360}$$

$$\text{pymt}\dfrac{(1.0075)^{360} - 1}{(0.0075)} = 100,000(1.0075)^{360}$$

$$\text{pymt}\dfrac{13.73057612}{0.0075} = 1,473,057.612$$

$$\text{pymt}(1830.74348) = 1,473,057.612$$

$$\text{pymt} = \dfrac{1,473,057.612}{1830.74348}$$

$$= 804.6226 \approx \$804.62$$

Exercise 5.4

17.d) Continued

$$\text{Interest} = \begin{pmatrix}\text{Number of} \\ \text{Payments}\end{pmatrix}\begin{pmatrix}\text{Payment}\end{pmatrix} - \text{Principal}$$

$$\text{Interest} = 360(804.62) - 100,000$$

$$= 289,663.20 - 100,000.00 = \$189,663.20$$

e) $i = \dfrac{10\%}{12} = \dfrac{0.10}{12}$

$$\text{pymt}\frac{(1 + i)^n - 1}{i} = P(1 + i)^n \qquad \text{Calculate pymt}$$

$$\text{pymt}\frac{\left(1 + \dfrac{0.10}{12}\right)^{360} - 1}{\left(\dfrac{0.10}{12}\right)} = 100,000\left(1 + \dfrac{0.10}{12}\right)^{360}$$

$$\text{pymt}\frac{(1.00833333)^{360} - 1}{(0.00833333)} = 100,000(1.00833333)^{360}$$

$$\text{pymt}\frac{18.8373994}{0.00833333} = 1,983,739.937$$

$$\text{pymt}(2260.48792) = 1,983,739.937$$

$$\text{pymt} = \frac{1,983,739.937}{2260.48792}$$

$$= 877.5715 \approx \$877.57$$

$$\text{Interest} = \begin{pmatrix}\text{Number of} \\ \text{Payments}\end{pmatrix}\begin{pmatrix}\text{Payment}\end{pmatrix} - \text{Principal}$$

$$\text{Interest} = 360(877.57) - 100,000$$

$$= 315,925.20 - 100,000.00 = \$215,925.20$$

f) $i = \dfrac{11\%}{12} = \dfrac{0.11}{12}$

$$\text{pymt}\frac{(1 + i)^n - 1}{i} = P(1 + i)^n \qquad \text{Calculate pymt}$$

$$\text{pymt}\frac{\left(1 + \dfrac{0.11}{12}\right)^{360} - 1}{\left(\dfrac{0.11}{12}\right)} = 100,000\left(1 + \dfrac{0.11}{12}\right)^{360}$$

$$\text{pymt}\frac{(1.009166667)^{360} - 1}{(0.009166667)} = 100,000(1.009166667)^{360}$$

17.f) Continued

$$\text{pymt}\frac{25.7080970}{0.00916667} = 2,670,809.70$$

$$\text{pymt}(2804.51967) = 2,670,809.70$$

$$\text{pymt} = \frac{2,670,809.70}{2804.51967}$$

$$= 952.32340 \approx \$952.32$$

$$\text{Interest} = \binom{\text{Number of}}{\text{Payments}}\binom{\text{Payment}}{} - \text{Principal}$$

$$\text{Interest} = 360(952.32) - 100,000$$

$$= 342,835.20 - 100,000.00 = \$242,835.20$$

Summary of Exercise 17.

	Interest Rate	Monthly Payment	Total Interest
a)	6.0%	$599.55	$115,838.00
b)	7.0%	665.30	139,508.00
c)	8.0%	733.76	164,153.60
d)	9.0%	804.62	189,663.20
e)	10.0%	877.57	215,925.20
f)	11.0%	952.32	242,835.20

21. $P = \$100,000$, $r = 10\% = 0.10$

a) **Monthly Payments:**

$$n = (30 \text{ years})(12 \text{ months}) = 360 \text{ months}$$

$$i = \frac{10\%}{12} = \frac{0.10}{12}$$

$$\text{pymt}\frac{(1 + i)^n - 1}{i} = P(1 + i)^n \qquad \text{Calculate pymt}$$

$$\text{pymt}\frac{\left(1 + \frac{0.10}{12}\right)^{360} - 1}{\left(\frac{0.10}{12}\right)} = 100,000\left(1 + \frac{0.10}{12}\right)^{360}$$

Exercise 5.4

21.a) Continued

$$\text{pymt}\frac{(1.008333333)^{360} - 1}{(0.008333333)} = 100,000(1.008333333)^{360}$$

$$\text{pymt}\frac{18.8373991}{0.00833333} = 1,983,739.91$$

$$\text{pymt}(2260.48790) = 1,983,739.91$$

$$\text{pymt} = \frac{1,983,739.91}{2260.48790}$$

$$= 877.5716 \approx \$877.57$$

Amount paid in 1 year = 12(877.57) = \$10,530.84

$$\text{Interest} = \binom{\text{Number of}}{\text{Payments}}\binom{\text{Payment}}{} - \text{Principal}$$

$$\text{Interest} = 360(877.57) - 100,000$$

$$= 315,925.20 - 100,000.00 = \$215,925.20$$

b) <u>Biweekly Payments:</u>
    n = (30 years)(26 periods) = 780 periods

$$i = \frac{10\%}{26} = \frac{0.10}{26}$$

$$\text{pymt}\frac{(1 + i)^n - 1}{i} = P(1 + i)^n \qquad \text{Calculate pymt}$$

$$\text{pymt}\frac{\left(1 + \frac{0.10}{26}\right)^{780} - 1}{\left(\frac{0.10}{26}\right)} = 100,000\left(1 + \frac{0.10}{26}\right)^{780}$$

$$\text{pymt}\frac{(1.003846154)^{780} - 1}{(0.003846154)} = 100,000(1.003846154)^{780}$$

$$\text{pymt}\frac{18.9702863}{0.003846154} = 1,997,028.63$$

$$\text{pymt}(4932.27444) = 1,997,028.63$$

$$\text{pymt} = \frac{1,997,028.63}{4932.27444}$$

$$= 404.8900 \approx \$404.89$$

Amount paid in 1 year = 26(404.89) = \$10,527.14

21.b) Continued

$$\text{Interest} = \left(\begin{array}{c}\text{Number of}\\\text{Payments}\end{array}\right)\left(\text{Payment}\right) - \text{Principal}$$

$$\text{Interest} = 780(404.89) - 100,000$$

$$= 315,814.20 - 100,000.00 = \$215,814.20$$

Biweekly payments are $3.70 per year less than monthly payments for a total interest savings of $111.00. This is not a significant savings for the home owner but it might be more convenient to make biweekly payments if the owner receives a biweekly salary.

25. $P = 48,000$, $i = \dfrac{9\frac{1}{4}\%}{12} = \dfrac{0.0925}{12}$, $n = 4$ months

a) <u>monthly payment</u>

$$\text{pymt}\frac{(1 + i)^n - 1}{i} = P(1 + i)^n \qquad \text{Calculate pymt}$$

$$\text{pymt}\frac{\left(1 + \dfrac{0.0925}{12}\right)^4 - 1}{\left(\dfrac{0.0925}{12}\right)} = 48,000\left(1 + \dfrac{0.0925}{12}\right)^4$$

$$\text{pymt}\frac{(1.007708333)^4 - 1}{(0.007708333)} = 48,000(1.007708333)^4$$

$$\text{pymt}\frac{0.03119168}{0.007708333} = 49,497.2006$$

$$\text{pymt}(4.04648811) = 49,497.2006$$

$$\text{pymt} = \frac{49,497.2006}{4.04648811}$$

$$= 12,232.1379 \approx \$12,232.14$$

b) <u>Calculations for Month 1 (months 2 and 3 are similar)</u>

$P = \$48,000$, $r = 9\frac{1}{4}\% = 0.0925$, $t = 1$ month $= \dfrac{1}{12}$

$I = Prt$ \qquad Calculate I

$$= (48,000)(0.0925)\left(\frac{1}{12}\right) = 370.00$$

Principal portion = payment - interest portion

$$= 12,232.14 - 370.00 = 11,862.14$$

Exercise 5.4

25.b) Continued

$$\begin{pmatrix} \text{Amount due} \\ \text{after Payment} \end{pmatrix} = \begin{pmatrix} \text{Previous} \\ \text{principal} \end{pmatrix} - \begin{pmatrix} \text{Principal} \\ \text{portion} \end{pmatrix}$$

$$= 48,000.00 - 11,862.14 = 36,137.86$$

<u>Calculations for Month 4 (the LAST month)</u>

P = 12,138.56

I = Prt   Calculate I

$$= (12,138.56)(0.0925)\left(\frac{1}{12}\right) = 93.57$$

Total payment = Principal portion + Interest portion

$$= 12,138.56 + 93.57 = 12,232.13$$

Payment Number	Principal Portion	Interest Portion	Total Payment	Amount Due After Payment
0	---	---	---	48,000.00
1	11,862.14	370.00	12,232.14	36,137.86
2	11,953.58	278.56	12,232.14	24,184.28
3	12,045.72	186.42	12,232.14	12,138.56
4	12,138.56	93.57	12,232.13	0.00

29. down payment = 10% of \$16,113.82 = \$1611.38

P = Cost - down payment = 16,113.82 - 1611.38 = \$14,502.44

n = 3 years, 2 months = (3 years)(12 months) + 2 months = 38

$$i = \frac{11\frac{1}{2}\%}{12} = \frac{0.115}{12}$$

Unpaid Balance $= P(1 + i)^n - \text{pymt}\dfrac{(1 + i)^n - 1}{i}$

$$= 14,502.44\left(1 + \frac{0.115}{12}\right)^{38} - 378.35\dfrac{\left(1 + \dfrac{0.115}{12}\right)^{38} - 1}{\left(\dfrac{0.115}{12}\right)}$$

29. Continued

$$= 14,502.44(1.00958333)^{38} - 378.35\left(\frac{(1.00958333)^{38} - 1}{0.00958333}\right)$$

$$= 20,837.4048 - 378.35\left(\frac{0.43682062}{0.00958333}\right)$$

$$= 20,837.4048 - 17,245.6781$$

$$= 3,591.7267 \approx \$3,591.73$$

33. a) <u>Monthly payment on existing loan</u>

P = \$152,850, n = (30 years)(12 months) = 360 months

$$i = \frac{13\frac{3}{8}\%}{12} = \frac{0.13375}{12}$$

$$\text{pymt}\frac{(1 + i)^n - 1}{i} = P(1 + i)^n \qquad \text{Calculate pymt}$$

$$\text{pymt}\frac{\left(1 + \frac{0.13375}{12}\right)^{360} - 1}{\left(\frac{0.13375}{12}\right)} = 152,850\left(1 + \frac{0.13375}{12}\right)^{360}$$

$$\text{pymt}\frac{(1.01114583)^{360} - 1}{(0.01114583)} = 152,850(1.01114583)^{360}$$

$$\text{pymt}\frac{53.0712943}{0.01114583} = 8,264,797.33$$

$$\text{pymt}(4761.53668) = 8,264,797.33$$

$$\text{pymt} = \frac{8,264,797.33}{4761.53668}$$

$$= 1735.7416 \approx \$1735.74$$

b) Find unpaid balance at end of 3 years on existing loan.

n = 3 years = (3 years)(12 months) = 36 months

$$i = \frac{13\frac{3}{8}\%}{12} = \frac{0.13375}{12}$$

Exercise 5.4

33.b) **Continued**

$$\text{Unpaid Balance} = P(1 + i)^n - \text{pymt}\frac{(1 + i)^n - 1}{i}$$

$$= 152,850\left(1 + \frac{0.13375}{12}\right)^{36} - 1735.74\,\frac{\left(1 + \frac{0.13375}{12}\right)^{36} - 1}{\left(\frac{0.13375}{12}\right)}$$

$$= 152,850(1.01114583)^{36} - 1735.74\left(\frac{(1.01114583)^{36} - 1}{0.01114583}\right)$$

$$= 227,804.406 - 1735.74\left(\frac{0.49037884}{0.01114583}\right)$$

$$= 227,804.406 - 76,366.670$$

$$= 151,437.736 = \$151,437.74$$

c) <u>**Monthly payment on new loan**</u>

$P = \$151,437.74$, $n = (30 \text{ years})(12 \text{ months}) = 360 \text{ months}$

$$i = \frac{8\frac{7}{8}\%}{12} = \frac{0.08875}{12}$$

$$\text{pymt}\frac{(1 + i)^n - 1}{i} = P(1 + i)^n \qquad \text{Calculate pymt}$$

$$\text{pymt}\frac{\left(1 + \frac{0.08875}{12}\right)^{360} - 1}{\left(\frac{0.08875}{12}\right)} = 151,437.74\left(1 + \frac{0.08875}{12}\right)^{360}$$

$$\text{pymt}\frac{(1.00739583)^{360} - 1}{(0.00739583)} = 151,437.74(1.00739583)^{360}$$

$$\text{pymt}\frac{13.192343}{0.00739583} = 2,149,256.28$$

$$\text{pymt}(1783.75336) = 2,149,256.28$$

$$\text{pymt} = \frac{2,149,256.28}{1783.75336}$$

$$= 1204.9066 \approx \$1204.91$$

33. **Continued**

   d) Total interest if they do **not** refinance

   $n = (30 \text{ years})(12 \text{ months}) = (30)(12) = 360 \text{ months}$

   $\text{Interest} = \begin{pmatrix} \text{Number of} \\ \text{Payments} \end{pmatrix} \begin{pmatrix} \text{Payment} \end{pmatrix} - \text{Principal}$

   $\text{Interest} = 360(1735.74) - 152,850.00$
   $= 624,866.40 - 152,850.00 = \$472,016.40$

   e) Total interest if they **do** refinance

   $n = (30 \text{ years})(12 \text{ months}) = 360 \text{ months}$

   $\text{Interest} = \begin{pmatrix} \text{Number of} \\ \text{Payments} \end{pmatrix} \begin{pmatrix} \text{Payment} \end{pmatrix} - \text{Principal}$

   $\text{Interest} = 360(1204.91) + 36(1735.74) - 152,850.00$

   $= 433,767.60 + 62,486.64 - 152,850.00$
   $= \$343,404.24$

   f) The Wolfs should refinance their loan. They will save $128,612.16 in interest. Their payments will also decrease by $530.83 per month.

37. a) $P = 100,000$, $n = (30 \text{ years})(12 \text{ months}) = 360$, $i = \dfrac{0.075}{12}$

   $$\text{pymt} \frac{(1 + i)^n - 1}{i} = P(1 + i)^n \qquad \text{Calculate pymt}$$

   $$\text{pymt} \frac{\left(1 + \dfrac{0.075}{12}\right)^{360} - 1}{\left(\dfrac{0.075}{12}\right)} = 100,000\left(1 + \dfrac{0.075}{12}\right)^{360}$$

   $$\text{pymt} \frac{(1.00625)^{360} - 1}{(0.00625)} = 100,000(1.00625)^{360}$$

   $$\text{pymt} \frac{8.421533905}{0.00625} = 942,153.3905$$

   $$\text{pymt}(1347.4454248) = 942,153.3905$$

   $$\text{pymt} = \frac{942,153.3905}{1347.4454248}$$

   $$= 699.2145086 \approx \$699.21$$

Exercise 5.4

37. Continued

b) $P = 100{,}000$, $n = (30 \text{ years})(12 \text{ months}) = 360$, $i = \dfrac{0.05375}{12}$

$$\text{pymt}\frac{(1 + i)^n - 1}{i} = P(1 + i)^n \qquad \text{Calculate pymt}$$

$$\text{pymt}\frac{\left(1 + \dfrac{0.05375}{12}\right)^{360} - 1}{\left(\dfrac{0.05375}{12}\right)} = 100{,}000\left(1 + \dfrac{0.05375}{12}\right)^{360}$$

$$\text{pymt}\frac{(1.004479167)^{360} - 1}{(0.004479167)} = 100{,}000(1.004479167)^{360}$$

$$\text{pymt}\frac{3.997308261}{0.004479167} = 499{,}730.8261$$

$$\text{pymt}(892.4223094) = 499{,}730.8261$$

$$\text{pymt} = \frac{499{,}730.8261}{892.4223094}$$

$$= 559.971239 \approx \$559.97$$

c)   Savings = Fixed rate payments – Adjustable rate payments
       = 699.21(12) – 559.97(12)
       = 8390.52 – 6719.64 = \$1670.88

d) $n = 12$

Unpaid Balance $= P(1 + i)^n - \text{pymt}\dfrac{(1 + i)^n - 1}{i}$

$$= 100{,}000\left(1 + \frac{0.05375}{12}\right)^{12} - 559.97\frac{\left(1 + \dfrac{0.05375}{12}\right)^{12} - 1}{\left(\dfrac{0.05375}{12}\right)}$$

$$= 100{,}000(1.004479167)^{12} - 559.97\left(\frac{(1.004479167)^{12} - 1}{0.004479167}\right)$$

$$= 105{,}509.4125 - 559.97\left(\frac{0.055094125}{0.004479167}\right)$$

$$= 105{,}509.4125 - 6887.6778$$

$$= 98{,}621.7347 \approx \$98{,}621.73$$

37. Continued

e) Second year rate = 5.375% + 2% = 7.375%   (adjust upward 2%)

   n = 360 months − 12 months = 348 months

f) P = 98,621.73, n = (29 years)(12 months) = 348, i = $\dfrac{0.07375}{12}$

$$\text{pymt}\frac{(1 + i)^n - 1}{i} = P(1 + i)^n \qquad \text{Calculate pymt}$$

$$\text{pymt}\frac{\left(1 + \dfrac{0.07375}{12}\right)^{348} - 1}{\left(\dfrac{0.07375}{12}\right)} = 98,621.73\left(1 + \dfrac{0.07375}{12}\right)^{348}$$

$$\text{pymt}\frac{(1.006145833)^{348} - 1}{(0.006145833)} = 98,621.73(1.006145833)^{348}$$

$$\text{pymt}\frac{7.433438722}{0.006145833} = 98,621.73(8.433438722)$$

$$\text{pymt}(1209.5086735) = 831,720.3167$$

$$\text{pymt} = \frac{831,720.3167}{1209.5086735}$$

$$= 687.65139 \approx \$687.65$$

g) Savings = Fixed rate payments − Adjustable rate payments
   = 699.21(24) − ((559.97)(12) + (687.65)(12))
   = 16,781.04 − (6719.64 + 8251.80)
   = 16,781.04 − 14,971.44 = $1809.60

h) Omitted.

41. Omitted.

Exercise 5.4

45. P = 170,000, r = 10.875%, n = 360 months, pymt = 1602.91

  a) Follow the instructions in the textbook to construct the
      following table:

Payment Number	Principal Portion	Interest Portion	Total Payment	Balance
0				$170,000.00
1	$62.29	$1,540.63	$1,602.91	$169,937.72
2	$62.85	$1,540.06	$1,602.91	$169,874.87
3	$63.42	$1,539.49	$1,602.91	$169,811.45
4	$63.99	$1,538.92	$1,602.91	$169,747.45
5	$64.57	$1,538.34	$1,602.91	$169,682.88
6	$65.16	$1,537.75	$1,602.91	$169,617.72
7	$65.75	$1,537.16	$1,602.91	$169,551.97
8	$66.35	$1,536.56	$1,602.91	$169,485.63
9	$66.95	$1,535.96	$1,602.91	$169,418.68
10	$67.55	$1,535.36	$1,602.91	$169,351.13
11	$68.17	$1,534.74	$1,602.91	$169,282.96
12	$68.78	$1,534.13	$1,602.91	$169,214.18

First Year
Totals $18,449.10

  b) First year's interest = $18,449.10.

  c) n = 29 years = 29(12) = 348 months

$$\text{Unpaid Balance} = P(1 + i)^n - \text{pymt}\,\frac{(1 + i)^n - 1}{i}$$

$$= 170,000\left(1 + \frac{0.10875}{12}\right)^{348} - 1602.91\,\frac{\left(1 + \frac{0.10875}{12}\right)^{348} - 1}{\left(\frac{0.10875}{12}\right)}$$

$$= 170,000(1.0090625)^{348} - 1602.91\left(\frac{(1.0090625)^{348} - 1}{0.0090625}\right)$$

$$= 3,925,871.967 - 1602.91\left(\frac{22.0933645}{0.0090625}\right)$$

$$= 3,925,871.967 - 3,907,715.852$$

$$= 18,156.115 \approx \$18,156.11$$

45. Continued

d)

Payment Number	Principal Portion	Interest Portion	Total Payment	Balance
348				$18,156.11
349	$1,438.37	$164.54	$1,602.91	$16,717.74
350	$1,451.41	$151.50	$1,602.91	$15,266.33
351	$1,464.56	$138.35	$1,602.91	$13,801.78
352	$1,477.83	$125.08	$1,602.91	$12,323.94
353	$1,491.22	$111.69	$1,602.91	$10,832.72
354	$1,504.74	$98.17	$1,602.91	$9,327.98
355	$1,518.38	$84.53	$1,602.91	$7,809.61
356	$1,532.14	$70.77	$1,602.91	$6,277.47
357	$1,546.02	$56.89	$1,602.91	$4,731.45
358	$1,560.03	$42.88	$1,602.91	$3,171.42
359	$1,574.17	$28.74	$1,602.91	$1,597.25
360	$1,597.25	$14.48	$1,602.91	$0.00

Last Year Totals $1,087.63

e) Last year's interest = $1,087.63

49. a) P = $10,120, n = 5 years(12 months) = 60 months

$$i = \frac{7\frac{7}{8}\%}{12} = \frac{0.07875}{12}$$

$$pymt\frac{(1 + i)^n - 1}{i} = P(1 + i)^n \qquad \text{Calculate pymt}$$

$$pymt\frac{\left(1 + \frac{0.07875}{12}\right)^{60} - 1}{\left(\frac{0.07875}{12}\right)} = 10,120\left(1 + \frac{0.07875}{12}\right)^{60}$$

$$pymt\frac{(1.0065625)^{60} - 1}{(0.0065625)} = 10,120(1.0065625)^{60}$$

$$pymt\frac{0.48062402}{0.0065625} = 14,983.91506$$

49.a)   Continued

$$\text{pymt}(73.237946) = 14,983.91506$$

$$\text{pymt} = \frac{14983.91506}{73.237946}$$

$$= 204.59224 \approx \$204.59$$

Payment Number	Principal Portion	Interest Portion	Total Payment	Balance
0				$10,120.00
1	$138.18	$66.41	$204.59	$9,981.82
2	$139.08	$65.51	$204.59	$9,842.74
3	$140.00	$64.59	$204.59	$9,702.74
4	$140.92	$63.67	$204.59	$9,561.83
5	$141.84	$62.75	$204.59	$9,419.98
6	$142.77	$61.82	$204.59	$9,277.21
7	$143.71	$60.88	$204.59	$9,133.51
8	$144.65	$59.94	$204.59	$8,988.85
9	$145.60	$58.99	$204.59	$8,843.25
10	$146.56	$58.03	$204.59	$8,696.70
11	$147.52	$57.07	$204.59	$8,549.18
12	$148.49	$56.10	$204.59	$8,400.69

First Year
Totals $735.77

b)   First Year's Interest = $735.77

c)   n = 4 years(12 months) = 48 months

$$\text{Unpaid Balance} = P(1 + i)^n - \text{pymt}\frac{(1 + i)^n - 1}{i}$$

$$= 10,120\left(1 + \frac{0.07875}{12}\right)^{48} - 204.59\frac{\left(1 + \frac{0.07875}{12}\right)^{48} - 1}{\left(\frac{0.07875}{12}\right)}$$

$$= 10,120(1.0065625)^{48} - 204.59\left(\frac{(1.0065625)^{48} - 1}{0.0065625}\right)$$

$$= 13,852.76 - 11,499.12$$

$$= 2353.64 \approx \$2353.64$$

49. Continued

d)

Payment Number	Principal Portion	Interest Portion	Total Payment	Balance
48				$2,353.64
49	$189.14	$15.45	$204.59	$2,164.50
50	$190.39	$14.20	$204.59	$1,974.11
51	$191.63	$12.96	$204.59	$1,782.48
52	$192.89	$11.70	$204.59	$1,589.58
53	$194.16	$10.43	$204.59	$1,395.42
54	$195.43	$9.16	$204.59	$1,199.99
55	$196.72	$7.87	$204.59	$1,003.28
56	$198.01	$6.58	$204.59	$805.27
57	$199.31	$5.28	$204.59	$605.97
58	$200.61	$3.98	$204.59	$405.35
59	$201.93	$2.66	$204.59	$203.42
60	$203.42	$1.33	$204.75	$0.00

First Year
Totals $101.61

e)   Last year's interest = $101.61

Exercise 5.5

1.   down payment = 10% of $16,113.82 = $1611.38

P = Cost − down payment = 16,113.82 − 1611.38 = $14,502.44

$i = \dfrac{11\frac{1}{2}\%}{12} = \dfrac{0.115}{12}$, n = (4 years)(12 months) = 48 months

$$\text{pymt}\frac{(1 + i)^n - 1}{i} = P(1 + i)^n \qquad \text{Calculate pymt}$$

$$\text{pymt}\frac{\left(1 + \dfrac{0.115}{12}\right)^{48} - 1}{\left(\dfrac{0.115}{12}\right)} = 14,502.44\left(1 + \dfrac{0.115}{12}\right)^{48}$$

$$\text{pymt}\frac{(1.009583333)^{48} - 1}{(0.009583333)} = 14,502.44(1.009583333)^{48}$$

Exercise 5.5

1. Continued

$$\text{pymt}\frac{0.58060837}{0.00958333} = 22,922.68$$

$$\text{pymt}(60.5852206) = 22,922.68$$

$$\text{pymt} = \frac{22,922.68}{60.5852206}$$

$$= 378.3543 \approx \$378.35$$

A.P.R. Financed Amount = Loan Amount - Loan Fees
$$= 14,502.44 - 814.14 = \$13,688.30$$

$$\text{pymt}\frac{(1 + i)^n - 1}{i} = P(1 + i)^n$$

Substitute in formula to find the two equations to graph.

$$378.35\frac{(1 + i)^{48} - 1}{i} = 13,688.30(1 + i)^{48}$$

$$y1 = 378.35\frac{(1 + x)^{48} - 1}{x}, \quad y2 = 13,688.30(1 + x)^{48}$$

Estimate the range from the calculation for the payment above.

xmax = i = 0.009583 ≈ 0.02,
y should approximate the right side of the payment equation.
ymin = 20,000, ymax = 30,000

Graph and find the intersection.

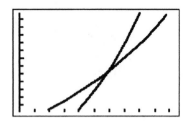

x = 0.01218645271, y = 24482.428273

Convert x to the annual rate.

annual rate = 12(0.01218645271) = 0.146237 ≈ 14.6%

This means that the 11.5% loan requires the same monthly payment as a 14.6% loan with no points or fees.

5.   P = Cost = $4600, n = (4 years)(12 months) = 48 months

   a)   <u>Monthly Payment:</u>   (Add-on Interest)   r = 8% = 0.08

         FV = P(1 + rt)
            = 4600(1 + (0.08)(4))
            = 4600(1.32) = 6072.00

         Monthly Payment = $\dfrac{6072.00}{48}$ = $126.50

   b)   Verify the A.P.R.

         $$\text{pymt}\dfrac{(1 + i)^n - 1}{i} = P(1 + i)^n$$

         Substitute in formula to find the two equations to graph.

         $$126.50\dfrac{(1 + i)^{48} - 1}{i} = 4600(1 + i)^{48}$$

         $$y1 = 126.50\dfrac{(1 + x)^{48} - 1}{x}, \quad y2 = 4600(1 + x)^{48}$$

         Estimate the range from the calculation for the payment
         above.

              xmax = i = 0.08/12 = 0.0066667 ≈ 0.02,
              y should approximate the future value in the payment
              calculation. ymin = 0, ymax = 10,000

   Graph and find the intersection.

   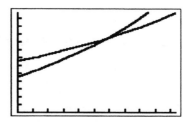

         x = 0.01195440844, y = 8137.3551821

   Convert x to the annual rate.

         annual rate = 12(0.01195440844) = 0.143453 ≈ 14.3%

   This means that the 8% add-on loan requires the same monthly
   payment as a 14.3% simple interest amortized loan would
   have. The A.P.R. is legally correct since 14.3453% - 14.25%
   = 0.095% which is less the 0.125%, the legal tolerance.

Exercise 5.5

9. It would seem that the savings and loan would be the less expensive loan, however, since the A.P.R. is not given it is not known what the effect of the finance charges would be and therefore it cannot be determined which loan would be least expensive.

13. $P = 80\%$ of $119,000 = 95,200$, $n = (30 \text{ years})(12 \text{ months}) = 360$,

$$i = \frac{8\frac{1}{4}\%}{12} = \frac{0.0825}{12}$$

Find the monthly payment:

$$\text{pymt}\frac{(1 + i)^n - 1}{i} = P(1 + i)^n \qquad \text{Calculate pymt}$$

$$\text{pymt}\frac{\left(1 + \dfrac{0.0825}{12}\right)^{360} - 1}{\left(\dfrac{0.0825}{12}\right)} = 95,200\left(1 + \frac{0.0825}{12}\right)^{360}$$

$$\text{pymt}\frac{(1.006875)^{360} - 1}{(0.006875)} = 95,200(1.006875)^{360}$$

$$\text{pymt}\frac{10.78150562}{0.006875} = 1,121,599.335$$

$$\text{pymt}(1568.218999) = 1,121,599.335$$

$$\text{pymt} = \frac{1,121,599.335}{1568.218999}$$

$$= 715.2058 \approx \$715.21$$

Approximate the fees included in the finance charge:

$$i = \frac{\text{A.P.R.}}{12} = \frac{9.23\%}{12} = \frac{0.0923}{12}$$

$$\text{pymt}\frac{(1 + i)^n - 1}{i} = P(1 + i)^n \qquad \text{Calculate P}$$

$$715.21\frac{\left(1 + \dfrac{0.0923}{12}\right)^{360} - 1}{\left(\dfrac{0.0923}{12}\right)} = P\left(1 + \frac{0.0923}{12}\right)^{360}$$

$$715.21\frac{(1.007691667)^{360} - 1}{(0.007691667)} = P(1.007691667)^{360}$$

$$715.21\frac{14.77466514}{0.007691667} = P(15.77466514)$$

13. Continued

$$715.21(1920.86654) = 15.77466514P$$

$$1,373,822.958 = 15.77466514P$$

$$P = \frac{1,373,822.958}{15.77466514}$$

$$= 87,090.4673 \approx \$87,090.47$$

Estimated points and fees = 95,200.00 − 87,090.47 = $8109.53

17. Omitted.

Exercise 5.6

1. pymt = 1200, n = (20 years)(12 months) = 240, $i = \frac{8\%}{12} = \frac{0.08}{12}$

    a) $$P(1 + i)^n = pymt\frac{(1 + i)^n - 1}{i}$$ Calculate P

    $$P\left(1 + \frac{0.08}{12}\right)^{240} = 1200\frac{\left(1 + \frac{0.08}{12}\right)^{240} - 1}{\left(\frac{0.08}{12}\right)}$$

    $$P(1.006666667)^{240} = 1200\frac{(1.006666667)^{240} - 1}{(0.006666667)}$$

    $$P(4.92680277) = 1200\frac{3.92680277}{0.006666667}$$

    $$4.92680277P = 706,824.499$$

    $$P = \frac{706,824.499}{4.92680277} = \$143,465.15$$

    b) Total payout = (20 years)(12 months)($1200) = $288,000

Exercise 5.6

5. $FV = \$143,465.15$, $i = \dfrac{8\%}{12} = \dfrac{0.08}{12}$,

n = (30 years)(12 months) = 360

$$\text{pymt}\frac{(1 + i)^n - 1}{i} = \text{Future Value} \qquad \text{Calculate pymt}$$

$$\text{pymt}\frac{\left(1 + \dfrac{0.08}{12}\right)^{360} - 1}{\left(\dfrac{0.08}{12}\right)} = 143,465.15$$

$$\text{pymt}\frac{(1.00666667)^{360} - 1}{(0.00666667)} = 143,465.15$$

$$\text{pymt}\frac{9.9357297}{0.00666667} = 143,465.15$$

$$\text{pymt}(1490.35945) = 143,465.15$$

$$\text{pymt} = \frac{143,465.15}{1490.35945}$$

$$= 96.2621 \approx \$96.26$$

b) Total paid in = (30 years)(12 months)($96.26)
    = $34,653.60

She receives $288,000.00 - $34,653.60 = $253,346.40 more than she paid into the annuity.

9. t = 20 years, pymt = $14,000, c = 4% = 0.04

a) r = 8% = 0.08

$$P = \text{pymt}\frac{1 - \left(\dfrac{1 + c}{1 + r}\right)^t}{r - c} \qquad \text{Calculate P}$$

$$P = 14,000\frac{1 - \left(\dfrac{1 + 0.04}{1 + 0.08}\right)^{20}}{0.08 - 0.04}$$

$$P = 14,000\frac{1 - \left(\dfrac{1.04}{1.08}\right)^{20}}{0.04} = 14,000\frac{1 - (0.96296296)^{20}}{0.04}$$

$$= 14,000\frac{1 - 0.47010154}{0.04} = 14,000\frac{0.52989846}{0.04}$$

$$= 14,000(13.247461) = 185,464.4601 \approx \$185,464.46$$

9. Continued

    b) First annual payout = $14,000 (there is no C.O.L.A.
        adjustment)

    c) Second annual payout = $P(1 + c)^{n - 1}$

        $$= 14,000(1 + 0.04)^{2 - 1}$$

        $$= 14,000(1.04)^{1} = \$14,560.00$$

    d) Last annual payout = $P(1 + c)^{n - 1}$

        $$= 14,000(1 + 0.04)^{20 - 1}$$

        $$= 14,000(1.04)^{19}$$

        $$= 14,000(2.10684918)$$

        $$= \$29,495.8885 \approx \$29,495.89$$

13. pymt = $1000 monthly for 25 years, c = 3% = 0.03,
    r = 10% compounded monthly

    a) $n = (1 \text{ year})(12 \text{ months}) = 12$

        $$FV = \text{pymt}\frac{(1 + i)^{n} - 1}{i} \qquad \text{Calculate ordinary FV}$$

        $$FV = 1000\frac{\left(1 + \dfrac{0.10}{12}\right)^{12} - 1}{\dfrac{0.10}{12}}$$

        $$= 1000(12.565568) = 12,565.5681 \approx \$12,565.57$$

    b) $$r = \left(1 + \frac{0.10}{12}\right)^{12} - 1 = 1.104713067 - 1 = 0.104713067$$

    c) $$P = \text{pymt}\frac{1 - \left(\dfrac{1 + c}{1 + r}\right)^{t}}{r - c} \qquad \text{Calculate P}$$

        $$P = 12,565.57\frac{1 - \left(\dfrac{1 + 0.03}{1 + 0.104713067}\right)^{25}}{0.104713067 - 0.03}$$

        $$P = 12,565.57\frac{1 - \left(\dfrac{1.03}{1.104713067}\right)^{25}}{0.074713067}$$

13.d) Continued

$$= 12{,}565.57\frac{1 - 0.17365742}{0.074713067} = 12{,}565.57\frac{0.82634258}{0.074713067}$$

$$= 12565.57(11.0602149) = 138{,}977.905 \approx \$138{,}977.91$$

d)   FV $= \$143{,}465.15$, $i = \frac{10\%}{12} = \frac{0.10}{12}$,

n $=$ (30 years)(12 months) $= 360$

$$\text{pymt}\frac{(1 + i)^n - 1}{i} = \text{Future Value} \qquad \text{Calculate pymt}$$

$$\text{pymt}\frac{\left(1 + \dfrac{0.10}{12}\right)^{360} - 1}{\left(\dfrac{0.10}{12}\right)} = 138{,}977.90$$

$$\text{pymt}\frac{(1.00833333)^{360} - 1}{(0.00833333)} = 138{,}977.90$$

$$\text{pymt}\frac{18.8373994}{0.00833333} = 138{,}977.90$$

$$\text{pymt}(2260.48793) = 138{,}977.90$$

$$\text{pymt} = \frac{138{,}977.90}{2260.48793}$$

$$= 61.4814 \approx \$61.48$$

17.   pymt $= 50{,}000$, n $= 20$ years, $i = 8\% = 0.08$

$$P(1 + i)^n = \text{pymt}\frac{(1 + i)^n - 1}{i} \qquad \text{Calculate P}$$

$$P(1 + 0.08)^{20} = 50{,}000\frac{(1 + 0.08)^{20}}{0.08}$$

$$P(1.08)^{20} = 50{,}000\frac{(1.08)^{20} - 1}{0.08}$$

$$P(4.66095714) = 50{,}000\frac{3.66095714}{0.08}$$

$$4.66095714P = 2{,}288{,}098.215$$

17. Continued

$$P = \frac{2,288,098.215}{4.66095714}$$

$$= 490,907.3704 \approx \$490,907.37$$

21. Omitted.

Chapter 5 Review

1. $P = \$8140$, $r = 9\frac{3}{4}\% = 0.0975$, $t = 11$ years

    $I = Prt \qquad$ Calculate I

    $= (8140)(0.0975)(11)$

    $= \$8730.15$

5. $P = \$7000$, $FV = $ Amount needed $= \$8000$, $r = 7\frac{1}{2}\% = 0.075$

    $FV = P(1 + rt) \qquad$ Calculate t

    $8000 = 7000(1 + (0.075)t)$

    $\frac{8000}{7000} = 1 + 0.075t$

    $1.14285714 = 1 + 0.075t$

    $0.14285714 = 0.075t$

    $\frac{0.14285714}{0.075} = t$

    $t = 1.904762 \approx 1.9$ years

9. $FV = \$250,000$ at age 65

    a) $i = \frac{10\frac{1}{8}\%}{4} = \frac{0.10125}{4}$,

    $n = (65 - 25 \text{ years})(4 \text{ quarters}) = 40(4) = 160$ quarters

    $FV = P(1 + i)^n \qquad$ Calculate P

    $250,000 = P\left(1 + \frac{0.10125}{4}\right)^{160}$

    $250,000 = P(1.0253125)^{160}$

9.a)   Continued

$$250,000 = P(54.5758277)$$

$$P = \frac{250,000}{54.5758277} = 4580.7826 \approx \$4580.78$$

b) $P = \$250,000$, $r = 10\frac{1}{8}\% = 0.10125$, $t = 1$ month $= \frac{1}{12}$ year

$I = Prt$ Calculate I

$= 250,000(0.10125)\left(\frac{1}{12}\right)$

$= 2109.375 \approx \$2109.38$

13. a) $I = \$1300$, $r = 8\frac{1}{4}\% = 0.0825$, $t = 1$ month $= \frac{1}{12}$ year

$I = Prt$ Calculate P

$1300 = P(0.0825)\left(\frac{1}{12}\right)$

$1300 = P(0.006875)$

$P = \frac{1300}{0.006875} = 189,090.9091 \approx \$189,090.91$

b) $FV = \$189,090.91$, $i = \frac{9\frac{3}{4}\%}{12} = \frac{0.0975}{12}$,

$n = (30\text{ years})(12\text{ months}) = 30(12) = 360$ months

$$FV = pymt\frac{(1 + i)^n - 1}{i} \qquad \text{Calculate ordinary pymt}$$

$$189,090.91 = pymt\frac{\left(1 + \frac{0.0975}{12}\right)^{360} - 1}{\frac{0.0975}{12}}$$

$$189,090.91 = pymt\frac{(1.008125)^{360} - 1}{0.008125}$$

$$189,090.91 = pymt\left(\frac{17.4152875}{0.008125}\right)$$

$$189,090.91 = pymt(2143.42000)$$

$$pymt = \frac{189,090.91}{2143.42000} = 88.219252 \approx \$88.22$$

17. pymt = 1700, n = (25 years)(12 months) = 300,

a) $i = \dfrac{6.1\%}{12} = \dfrac{0.061}{12}$

$$P(1 + i)^n = \text{pymt} \frac{(1 + i)^n - 1}{i} \qquad \text{Calculate P}$$

$$P\left(1 + \frac{0.061}{12}\right)^{300} = 1700\frac{\left(1 + \dfrac{0.061}{12}\right)^{300} - 1}{\left(\dfrac{0.061}{12}\right)}$$

$$P(1.005083333)^{300} = 1700\frac{(1.005083333)^{300} - 1}{(0.005083333)}$$

$$P(4.57742697) = 1700\frac{3.57742697}{0.005083333}$$

$$4.57742697P = 1,196,385.414$$

$$P = \frac{1,196,385.414}{4.57742697}$$

$$= 261,366.357 \approx \$261,366.36$$

b) FV = $261,366.36, $i = \dfrac{6.1\%}{12} = \dfrac{0.061}{12}$,

n = (30 years)(12 months) = 360

$$\text{pymt}\frac{(1 + i)^n - 1}{i} = \text{Future Value} \qquad \text{Calculate pymt}$$

$$\text{pymt}\frac{\left(1 + \dfrac{0.061}{12}\right)^{360} - 1}{\left(\dfrac{0.061}{12}\right)} = 261,366.36$$

$$\text{pymt}\frac{(1.00508333)^{360} - 1}{(0.00508333)} = 261,366.36$$

$$\text{pymt}\frac{5.2050561}{0.00508333} = 261,366.36$$

$$\text{pymt}(1023.94545) = 261,366.36$$

$$\text{pymt} = \frac{261,366.36}{1023.94545}$$

$$= 255.2542 \approx \$255.25$$

c) Total paid in = (30 years)(12 months)($255.25) = $91,890.00
    Total received = (25 years)(12 months)($1700) = $510,000.00

6   Geometry

1.   Triangle: b = 9.2 cm, h = 3.5 cm

$A = \frac{1}{2}bh = \frac{1}{2}(9.2 \text{ cm})(3.5\text{cm}) = 16.1 \text{ cm}^2$

5.   Parallelogram: b = 6.2 ft, h = 3.5 ft

$A = bh = (6.2 \text{ ft})(3.5 \text{ ft}) = 21.7 \text{ ft}^2$

9.   Circle: d = 6.5 in., $r = \frac{d}{2} = \frac{6.5 \text{ in.}}{2} = 3.25$ in.

   a) $A = \pi r^2 = \pi(3.25 \text{ in.})^2 = \pi(10.5625 \text{ in.}^2)$

            $= 33.1831 \text{ in}^2 \approx 33.2 \text{ in.}^2$

   b) $C = \pi d = \pi(6.5 \text{ in.}) = 20.4204 \text{ in.} \approx 20.4 \text{ in.}$

13. Triangle: a = 8 m, b = 8 m, c = 8 m

   a)   Use Heron's Formula

            $s = \frac{1}{2}(a + b + c) = \frac{1}{2}(8 \text{ m} + 8 \text{ m} + 8 \text{ m}) = \frac{1}{2}(24 \text{ m}) = 12 \text{ m}$

            $A = \sqrt{s(s - a)(s - b)(s - c)}$

            $A = \sqrt{12(12 - 8)(12 - 8)(12 - 8)}$

            $A = \sqrt{12 \text{ m}(4 \text{ m})(4 \text{ m})(4 \text{ m})}$

            $A = \sqrt{768 \text{ m}^4} = 27.7128 \text{ m}^2 \approx 27.7 \text{ m}^2$

   b) $P = a + b + c = 8 \text{ m} + 8 \text{ m} + 8 \text{ m} = 24 \text{ m}$

17. Semicircle: $d = 100$ yd, $r = \dfrac{d}{2} = \dfrac{100}{2} = 50$ yd

   a) $A = \dfrac{1}{2}(\pi r^2) = \dfrac{1}{2}\pi(50 \text{ yd})^2 = \dfrac{2500}{2}\pi \text{ yd}^2$

   $= 3926.99082 \text{ yd}^2 \approx 3927.0 \text{ yd}^2$

   b) $C = \dfrac{1}{2}\pi d + d = \dfrac{1}{2}\pi(100 \text{ yd}) + 100 \text{ yd} = (50\pi + 100) \text{ yd}$

   $= (157.0796 + 100.00) \text{ yd} = 257.0796 \text{ yd} \approx 257.1 \text{ yd}$

21. Norman window: $b = d = 5$ ft, $h = 8$ ft, $r = \dfrac{d}{2} = \dfrac{5}{2} = 2.5$ ft

   a) $A_{total} = A_{rectangle} + A_{semicircle}$

   $= bh + \dfrac{1}{2}(\pi r^2) = (5 \text{ ft})(8 \text{ ft}) + \dfrac{1}{2}\pi(2.5 \text{ ft })^2$

   $= 40 \text{ ft}^2 + \dfrac{6.25}{2}\pi \text{ ft}^2 = (40 + 3.125\pi) \text{ ft}^2$

   $= (40 + 9.8175) \text{ ft}^2 = 49.8175 \text{ ft2} \approx 49.8 \text{ ft}^2$

   b) $P = \dfrac{1}{2}\text{Circumference} + 1 \text{ base} + 2 \text{ heights}$

   $= \dfrac{1}{2}(\pi d) + b + 2h = \dfrac{1}{2}\pi(5 \text{ ft}) + 5 \text{ ft} + 2(8 \text{ ft})$

   $= (7.8540 + 5 + 16) \text{ ft} = 28.8540 \text{ ft} \approx 28.9 \text{ ft}$

25. Square: $P = 18$ ft 8 in.   (Need to find length of sides $= s$)

   a) Area in square inches

   $P = 18 \text{ ft}\left(\dfrac{12 \text{ in.}}{1 \text{ ft}}\right) + 8 \text{ in.} = (216 + 8) \text{ in.} = 224 \text{ in.}$

   $s = \dfrac{P}{4} = \dfrac{224 \text{ in.}}{4} = 56 \text{ in.}$

   $A = s^2 = (56 \text{ in.})^2 = 3{,}136 \text{ in.}^2$

   b) Area in square feet

   $P = 18 \text{ ft} + 8 \text{ in.}\left(\dfrac{1 \text{ ft}}{12 \text{ in.}}\right) = (18 + \dfrac{8}{12}) \text{ ft} = 18.6667 \text{ ft}$

   $s = \dfrac{18.6667}{4} = 4.66667 \text{ ft}$

   $A = s^2 = (4.6667 \text{ ft})^2 = 21.7778 \text{ ft}^2 \approx 21.8 \text{ ft}^2$

Exercise 6.1

29. Triangle: Find the hypotenuse using the Pythagorean Theorem.

$a = \dfrac{3}{4}$ mi, $b = 1\dfrac{1}{2}$ mi

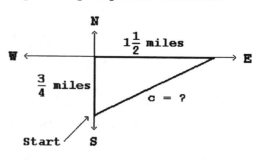

$c^2 = a^2 + b^2$

$= \left(\dfrac{3}{4} \text{ mi}\right)^2 + \left(1\dfrac{1}{2} \text{ mi}\right)^2$

$= \left(\dfrac{9}{16} + \dfrac{9}{4}\right) \text{ mi}^2$

$= \dfrac{45}{16} \text{ mi}^2 = 2.8125 \text{ mi}^2$

$c = \sqrt{2.8125 \text{ mi}^2} = 1.67705 \text{ mi} \approx 1.68 \text{ mi}$

Miles jogged $= P = a + b + c = \left(\dfrac{3}{4} + 1\dfrac{1}{2} + 1.677\right) \text{ mi}$

$\phantom{\text{Miles jogged} = P} = (0.75 + 1.50 + 1.677) \text{ mi} = 3.927 \text{ mi} \approx 3.9 \text{ mi}$

33. Oval field: $b = 100$ yd, $h = d = 40$ yd, $r = \dfrac{d}{2} = \dfrac{40}{2} = 20$ yd

a) $A_{total} = A_{rectangle} + 2 \cdot A_{semicircle}$

$\phantom{A_{total}} = bh + 2\left(\dfrac{1}{2}\right)(\pi r^2)$

$\phantom{A_{total}} = (100 \text{ yd})(40 \text{ yd}) + \pi(20 \text{ yd})^2 = 4000 \text{ yd}^2 + 400\pi \text{ yd}^2$

$\phantom{A_{total}} = (4000 + 1256.6371) \text{ yd}^2 = 5256.6371 \text{ yd}^2 \approx 5256.6 \text{ yd}^2$

b) $P = 2(\text{Circumference of semicircle}) + 2 \text{ bases}$

$ = 2\left(\dfrac{1}{2}\right)(\pi d) + 2b$

$ = \pi(40 \text{ yd}) + 2(100 \text{ yd})$

$ = (125.6637 + 200) \text{ yd} = 325.6637 \text{ yd} \approx 325.7 \text{ yd}$

37. <u>Small</u> (13 in.) for \$11.75: $d = 13$ in., $r = \dfrac{d}{2} = \dfrac{13}{2} = 6.5$ in.

$A = \pi r^2 = \pi(6.5 \text{ in.})^2 = 132.7323 \text{ in.}^2$

price per area $= \dfrac{\$11.75}{132.7323 \text{ in.}^2} = \$0.0885$ per in.^{2}

<u>Large</u> (16 in.) for \$14.75: $d = 16$ in., $r = \dfrac{d}{2} = \dfrac{16}{2} = 8$ in.

$A = \pi r^2 = \pi(8 \text{ in.})^2 = 201.0619 \text{ in.}^2$

price per area $= \dfrac{\$14.75}{201.0619 \text{ in.}^2} = \$0.0734$ per in.^{2}

<u>Super</u> (19 in.) for \$22.75: $d = 19$ in., $r = \dfrac{d}{2} = \dfrac{19}{2} = 9.5$ in.

$A = \pi r^2 = \pi(9.5 \text{ in.})^2 = 283.5287 \text{ in.}^2$

price per area $= \dfrac{\$22.75}{283.5287 \text{ in.}^2} = \$0.0802$ per in.^{2}

The best deal is the large pizza for \$14.75 since it costs only \$0.0734 per square inch.

Exercise 6.2

1. Rectangular box: $L = 5.2$ m, $w = 2.1$ m, $h = 3.5$ m

a) $V = L \cdot w \cdot h = (5.2 \text{ m})(2.1 \text{ m})(3.5 \text{ m}) = 38.22 \text{ m}^3$

b) $A_{box} = 2 \cdot A_{base} + 2 \cdot A_{front} + 2 \cdot A_{side}$

$= 2 \cdot L \cdot w + 2 \cdot L \cdot h + 2 \cdot w \cdot h$

$= [2(5.2)(2.1) + 2(5.2)(3.5) + 2(2.1)(3.5)] \text{ m}^2$

$= (21.84 + 36.40 + 14.70) \text{ m}^2$

$= 72.94 \text{ m}^2$

Exercise 6.2

5. Sphere: $d = 1\frac{3}{4}$ in. $= 1.75$ in., $r = \frac{d}{2} = \frac{1.75}{2} = 0.875$ in.

    a) $V = \frac{4}{3}\pi r^3 = \frac{4}{3}\pi(0.875 \text{ in.})^3 = \frac{4}{3}\pi(0.6699 \text{ in.}^3)$

    $= 2.8062 \text{ in.}^3 \approx 2.81 \text{ in.}^3$

    b) $A = 4\pi r^2 = 4\pi(0.875 \text{ in.})^2 = 4\pi(0.7656 \text{ in.}^2)$

    $= 9.6211 \text{ in.}^2 \approx 9.62 \text{ in.}^2$

9. Pyramid with square base: $s = 4$ ft, $h = 4$ ft

    $V = \frac{1}{3}A_{base} \cdot h = \frac{1}{3}s^2 h = \frac{1}{3}(4 \text{ ft})^2(4 \text{ ft}) = \frac{64}{3} \text{ ft}^3$

    $= 21.3333 \text{ ft}^3 \approx 21.33 \text{ ft}^3$

13. Two cylinders: $h = 5$ ft, $r_{outer} = 2$ ft, $r_{inner} = 1$ ft

    $V_{total} = V_{outer} - V_{inner} = A_{outer\ base} \cdot h - A_{inner\ base} \cdot h$

    $= \pi(r_{outer})^2 \cdot h - \pi(r_{inner})^2 \cdot h$

    $= \pi(2 \text{ ft})^2(5 \text{ ft}) - \pi(1 \text{ ft})^2(5 \text{ ft})$

    $= 20\pi \text{ ft}^3 - 5\pi \text{ ft}^3 = (20\pi - 5\pi) \text{ ft}^3$

    $= 15\pi \text{ ft}^3 = 47.12389 \text{ ft}^3 \approx 47.12 \text{ ft}^3$

17. Silo: $h = 50$ ft, $d = 20$ ft, $r = \frac{20}{2} = 10$ ft

    $V_{silo} = V_{cylinder} + V_{hemisphere}$

    $= A_{base} \cdot h + \frac{1}{2}V_{sphere}$

    $= (\pi r^2)h + \frac{1}{2}\left(\frac{4}{3}\pi r^3\right)$

    $= \pi(10 \text{ ft})^2(50 \text{ ft}) + \frac{2}{3}\pi(10 \text{ ft})^3$

    $= 5000\pi \text{ ft}^3 + 666.67\pi \text{ ft}^3 = 5666.67\pi \text{ ft}^3$

    $= 17,802.358 \text{ ft}^3 \approx 17,802.36 \text{ ft}^3$

21. Earth: d = 7920 mi, $r = \dfrac{7920}{2} = 3960$ mi

$$V_{earth} = \frac{4}{3}\pi r^3 = \frac{4}{3}\pi (3960 \text{ mi})^3 = 2.601203 \times 10^{11} \text{ mi}^3$$

Moon: d = 2160 mi, $r = \dfrac{2160}{2}$ mi $= 1080$ mi

$$V_{moon} = \frac{4}{3}\pi r^3 = \frac{4}{3}\pi (1080 \text{ mi})^3 = 5.276669 \times 10^9 \text{ mi}^3$$

$$\frac{V_{earth}}{V_{moon}} = \frac{2.601203 \times 10^{11} \text{ mi}^3}{5.276669 \times 10^9 \text{ mi}^3} = 49.296$$

Alternate Method:

$$\frac{V_{earth}}{V_{moon}} = \frac{\frac{4}{3}\pi (3960 \text{ mi})^3}{\frac{4}{3}\pi (1080 \text{ mi})^3} = \frac{(3960 \text{ mi})^3}{(1080 \text{ mi})^3} = 49.296$$

Approximately 49 moons could fit inside the earth.

25. Volume of one cord of wood = 128 ft³,
price of one cord = $190
purchased: L = 10 ft, w = 4 ft, h = 2 ft

$V_{purchased} = L \cdot w \cdot h = (10 \text{ ft})(4 \text{ ft})(2 \text{ ft}) = 80 \text{ ft}^3$

You did not get a cord of wood so you did not get an honest deal.
You received only 80/128 = 0.625 of a cord so you should have
paid (0.625)($190) = $118.75.

29. Cement driveway: $L = 36 \text{ ft}\left(\dfrac{1 \text{ yd}}{3 \text{ ft}}\right) = 12$ yd,

$$w = 9 \text{ ft}\left(\frac{1 \text{ yd}}{3 \text{ ft}}\right) = 3 \text{ yd}, \quad h = 6 \text{ in.}\left(\frac{1 \text{ yd}}{36 \text{ in.}}\right) = \frac{1}{6} \text{ yd}$$

$$V_{driveway} = L \cdot w \cdot h = (12 \text{ yd})(3 \text{ yd})\left(\frac{1}{6}\text{yd}\right) = 6 \text{ yd}^3$$

4 sacks of dry cement mix = 1 cubic yard of cement
6 cubic yards of cement require (6)(4) = 24 sacks of mix
24 sacks at $7.30 = 24($7.30) = $175.20
It will cost Ron $175.20 to buy the cement for his driveway.

Exercise 6.2

33. Water tank in shape of cone: d = 12 ft, r = $\frac{12}{2}$ = 6 ft

    Use the Pythagorean Theorem to find the height of the cone.
    (a = 6 ft, c = 10 ft, b = h)

    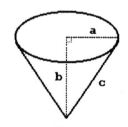

    $$c^2 = a^2 + b^2$$

    $$10^2 = 6^2 + h^2$$

    $$100 = 36 + h^2$$

    $$h^2 = 100 - 36 = 64, \quad \text{so } h = \sqrt{64} = 8 \text{ ft}$$

    $$V_{tank} = \frac{1}{3}A_{top} \cdot h = \frac{1}{3}(\pi r^2)h = \frac{1}{3}\pi(6 \text{ ft})^2(8 \text{ ft}) = 96\pi \text{ ft}^3$$

    $$= 301.59289 \text{ ft}^3 \approx 301.59 \text{ ft}^3$$

Exercise 6.3

1. Length of segments for the sides are 6, 8 and 10 units.
   The largest value = c = 10.

   $$a^2 + b^2 = 6^2 + 8^2 = 36 + 64 = 100 = 10^2 = c^2$$

   The segment lengths satisfy the Pythagorean Theorem, therefore
   the triangle is a right triangle with the right angle opposite
   the side of length 10.

5. a)   Truncated pyramid: h = 12 cubits, a = 2 cubits,
                            b = 15 cubits

   $$V = \frac{h}{3}(a^2 + ab + b^2)$$

   $$V = \frac{12 \text{ cubits}}{3}[(2 \text{ cubits})^2 + (2 \text{ cubits})(15 \text{ cubits}) + (15 \text{ cubits})^2]$$

   $$= (4 \text{ cubits})(4 \text{ cubits}^2 + 30 \text{ cubits}^2 + 225 \text{ cubits}^2)$$

5.a) Continued

$$= (4 \text{ cubits})(259 \text{ cubits}^2)$$

$$= 1036 \text{ cubits}^3 \left(\frac{\frac{3}{2} \text{ khar}}{1 \text{ cubits}^3}\right) = 1554 \text{ khar}$$

b)   Regular pyramid: h = 18 cubits, b = 15 cubits

$$V = \frac{h}{3}(b^2) = \frac{18 \text{ cubits}}{3}(15 \text{ cubits})^2 = (6 \text{ cubits})(225 \text{ cubits}^2)$$

$$= 1350 \text{ cubits}^3 \left(\frac{\frac{3}{2} \text{ khar}}{1 \text{ cubits}^3}\right) = 2025 \text{ khar}$$

The regular pyramid has the larger storage capacity.

9.   Construct a square whose sides are 24 units each, and inscribe a circle in the square. Thus, the diameter of the circle is 24 units. Now divide the sides of the square into three equal segments of 8 units each. Remove the triangular corners as shown and an irregular octagon is formed. The area of this octagon is then used as an approximation of the area of the circle:

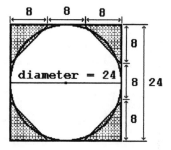

$$A_{circle} \approx A_{octagon}.$$

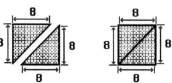

The area of the octagon is equal to the area of the large square minus the area of the four triangular corners. Rearranging these triangles, we can see that the area of two triangular corners is the same as the area of one square of side 8 units.

$$A_{circle} \approx A_{octagon} = A_{square \text{ of side } 24} - 2A_{square \text{ of side } 8}$$

$$= (24)^2 - 2(8)^2 = 576 - 128 = 448 \text{ square units}$$

$$\sqrt{448} = 21.1660105 \approx 21 \text{ units. Therefore, we can say}$$

$$A_{circle \text{ of diameter } 24} \approx 448 \approx 441 = A_{square \text{ of side } 21}$$

Exercise 6.3

13. Circle: r = 3 palms

a) Egyptian approximation of pi ($\pi = \frac{256}{81}$)

$A = \pi r^2 = \frac{256}{81}(3 \text{ palms})^2 = \frac{256}{81}(9 \text{ palms}^2) = \frac{256}{9} \text{ palms}^2$

$= 28.44444444 \text{ palms}^2 \approx 28.4444 \text{ palms}^2$

b) Calculator value of pi.

$A = \pi r^2 = \pi(3 \text{ palms})^2 = 9\pi \text{ palms}^2 = 28.27433388 \text{ palms}^2$

$\approx 28.2743 \text{ palms}^2$

c) Error of Egyptian calculation relative to the calculator value.

$A_{Egyptian} - A_{Calculator} = 28.44444444 - 28.27433388$

$= 0.17011056 \text{ palms}^2$

$\text{relative error} = \frac{0.17011056}{28.27433388} = 0.0060164 \approx 0.6\%$

17. a = 2 cubits, 5 palms, 3 fingers
b = 3 cubits, 3 palms, 2 fingers
c = 4 cubits, 4 palms, 3 fingers

```
P = a + b + c =   2 cubits,  5 palms, 3 fingers
                + 3 cubits,  3 palms, 2 fingers
                + 4 cubits,  4 palms, 3 fingers

                  9 cubits, 12 palms, 8 fingers
```

Convert fingers to palms (1 palm = 4 fingers):

```
                  9 cubits, 12 palms,  8 fingers
                           +2 palms, -8 fingers
                  9 cubits, 14 palms
```

Convert palms to cubits (1 cubit = 7 palms):

```
                  9 cubits,  14 palms
                 +2 cubits, -14 palms
                  11 cubits
```

The perimeter of the triangle is 11 cubits.

21. Triangle: a = 150 cubits, b = 200 cubits, c = 250 cubits

Use Heron's Formula

$$s = \frac{1}{2}(a + b + c) = \frac{1}{2}(150 + 200 + 250) = 300 \text{ cubits}$$

$$A = \sqrt{s(s - a)(s - b)(s - c)}$$

$$= \sqrt{300(300 - 150)(300 - 200)(300 - 250)\text{cubits}^4}$$

$$= \sqrt{(300)(150)(100)(50)\text{ cubits}^4}$$

$$= \sqrt{225,000,000 \text{ cubits}^4}$$

$$= 1500 \text{ cubits}^2 \left(\frac{1 \text{ setat}}{10,000 \text{ cubits}^2}\right) = 1.5 \text{ setats}$$

25. a)   d = 9 cubits, h = 10 cubits

Construct a square whose sides
are 9 units each, and inscribe a
circle in the square. Thus, the
diameter of the circle is 9
units. Now divide the sides of
the square into three equal
segments of 3 units each.
Remove the triangular corners as
shown and an irregular octagon
is formed. The area of this
octagon is then used as an
approximation of the area of the
circle:

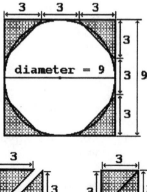

$$A_{circle} \approx A_{octagon}.$$

The area of the octagon is equal
to the area of the large square
minus the area of the four triangular corners. Rearranging
these triangles, we can see that the area of two triangular
corners is the same as the area of one square of side 3
units.

195

Exercise 6.3

25.a)   Continued

$$A_{circle} \approx A_{octagon} = A_{square \ of \ side \ 9} - 2A_{square \ of \ side \ 3}$$

$$= (9)^2 - 2(3)^2 = 81 - 18 = 63 \text{ square units}$$

$\sqrt{63} = 7.9372539 \approx 8$ units.   Therefore, we can say

$$A_{circle \ of \ diameter \ 9} \approx 63 \approx 64 = A_{square \ of \ side \ 8}$$

The eqyptians thought the area of the circle must be exactly the area of the square so they would use 64 square cubits as the area of the circle.

$$V = A_{circle} \cdot h = (64 \text{ cubits}^2)(10 \text{ cubits})$$

$$= 640 \text{ cubits}^3 \left(\frac{\frac{3}{2} \text{ khar}}{1 \text{ cubits}^3} \right) = 960 \text{ khar}$$

b)   r = 4.5 cubits, h = 10 cubits

$$V = A_{circle} \cdot h = \pi r^2 h = \pi(4.5)^2(10)$$

$$= 636.1725124 \text{ cubits}^3 \left(\frac{\frac{3}{2} \text{ khar}}{1 \text{ cubits}^3} \right)$$

$$= 954.2587685 \text{ khar} \approx 954.2588 \text{ khar}$$

c)   Error of Egyptian calculation relative to the calculator value.

$$V_{Egyptian} - V_{Calculator} = 960 - 954.2587685$$

$$= 5.74123147 \text{ khar}$$

$$\text{relative error} = \frac{5.74123147}{954.2587685} = 0.0060164 \approx 0.6\%$$

29.   Omitted.

Exercise 6.4

1. The sides are proportional because the triangles are similar.

    Compare the large triangle to the small triangle to find y:

    $$\frac{y}{4} = \frac{90}{6}, \text{ then}$$

    $$y = \frac{90(4)}{6} = \frac{360}{6} = 60$$

    Compare the small triangle to the large triangle to find x:

    $$\frac{x}{75} = \frac{6}{90}, \text{ then}$$

    $$x = \frac{6(75)}{90} = \frac{450}{90} = 5$$

5. Use similar triangles.

    x = height of tree

    $$\frac{x}{6} = \frac{21}{3.5}$$

    $$x = \frac{21(6)}{3.5} = \frac{126}{3.5} = 36$$

    The tree is 36 feet tall.

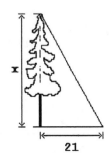

Exercise 6.4

9.   Rectangular Box:
          L = 4 ft,
          w = 3 ft,
          h = 2 ft

      c = diagonal on bottom of box
      d = diagonal of the box

      $c^2 = L^2 + w^2$

      $c^2 = (4 \text{ ft})^2 + (3 \text{ ft})^2$

          $= 16 \text{ ft}^2 + 9 \text{ ft}^2 = 25 \text{ ft}^2$

      $c = \sqrt{25 \text{ ft}^2} = 5 \text{ ft}$

      $d^2 = c^2 + h^2 = (5 \text{ ft})^2 + (2 \text{ ft})^2 = 25 \text{ ft}^2 + 4 \text{ ft}^2 = 29 \text{ ft}^2$

      $d = \sqrt{29 \text{ ft}^2} = \sqrt{29} \text{ ft} = 5.385 \text{ ft} \approx 5.3 \text{ ft}$

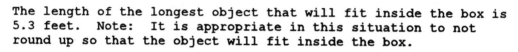

      The length of the longest object that will fit inside the box is
      5.3 feet. Note: It is appropriate in this situation to not
      round up so that the object will fit inside the box.

13.   Given: AD = CD
               AB = CB

      Prove: ∠DBA = ∠DBC

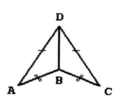

Statements	Reasons
1. AD = CD	1. Given
2. AB = CB	2. Given
3. DB = DB	3. Anything equals itself
4. ΔADB ≅ ΔCDB	4. SSS
5. ∠DBA = ∠DBC	5. Corresponding parts of congruent triangles are equal

17. Given: AE = CE
AB = CB

Prove: ∠ADB = ∠CDB

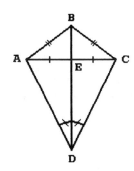

Statements	Reasons
1. AE = CE	1. Given
2. AB = CB	2. Given
3. BE = BE	3. Anything equals itself
4. △ABE ≅ △CBE	4. SSS
5. ∠ABD = ∠CBD	5. Corresponding parts of congruent triangles are equal
6. BD = BD	6. Anything equals itself
7. △ABD ≅ △CBD	7. SAS
8. ∠ADB = ∠CDB	8. Corresponding parts of congruent triangles are equal

21. a) $s = \sqrt{2 - \sqrt{2 + \sqrt{2}}}\ r = \sqrt{2 - \sqrt{2 + 1.41421356}}\ r$

$= \sqrt{2 - \sqrt{3.41421356}}\ r = \sqrt{2 - 1.84775907}\ r$

$= \sqrt{0.15224094}\ r = 0.39018064r$

$C_{circle} \approx P_{inscribed\ polygon\ with\ 16\ sides}$

$2\pi r \approx 16s$

$2\pi r \approx 16(0.39018064r)$

Exercise 6.4

21.a)   Continued

$$\pi = \frac{16(0.39018064)r}{2r} = 8(0.39018064) = 3.121445152$$

b) $s = \dfrac{2\sqrt{2 - \sqrt{2}}}{2 + \sqrt{2 + \sqrt{2}}} \; r = \dfrac{2\sqrt{2 - 1.41421356}}{2 + \sqrt{2 + 1.41421356}} \; r$

$= \dfrac{2\sqrt{0.58578644}}{2 + \sqrt{3.41421356}} \; r = \dfrac{2(0.76536686)}{2 + 1.84775907} \; r = \dfrac{1.53073373}{3.84775907} \; r$

$= 0.39782473r$

$C_{circle} \approx P_{circumscribed \; polygon \; with \; 16 \; sides}$

$2\pi r \approx 16s$

$2\pi r \approx 16(0.39782473r)$

$$\pi = \frac{16(0.39782473)r}{2r} = 8(0.39782473) = 3.182597878$$

Combining b) with a), we have $3.121445152 < \pi < 3.182597878$

25.   Omitted.

29.   Omitted.

Exercise 6.5

1.   opposite = x, adjacent = y, hypotenuse = 6

$\theta + 30° + 90° = 180°$ Sum of angles is 180°.

$\theta = 180° - 30° - 90° = 60°$

$\sin 30° = \dfrac{opp}{hyp} = \dfrac{x}{6},$ then $x = 6(\sin 30°) = 6\left(\dfrac{1}{2}\right) = 3$

$\cos 30° = \dfrac{adj}{hyp} = \dfrac{y}{6},$ then $y = 6(\cos 30°) = 6\left(\dfrac{\sqrt{3}}{2}\right) = 3\sqrt{3}$

5. x = opp, y = hyp, 3 = adj

    θ + 45° + 90° = 180°   Sum of angles is 180°.
         θ = 180° − 45° − 90° = 45°

    $\tan 45° = \dfrac{opp}{adj} = \dfrac{x}{3}$,   then x = 3(tan 45°) = 3(1) = 3

    $\cos 45° = \dfrac{adj}{hyp} = \dfrac{3}{y}$,   then $y = \dfrac{3}{\cos 45°} = \dfrac{3}{\left(\frac{1}{\sqrt{2}}\right)} = 3\sqrt{2}$

9. a = 12.0, A = 37°

    B = 180° − 90° − 37° = 53°

    $\tan 37° = \dfrac{opp}{adj} = \dfrac{12.0}{b}$

    b(tan 37°) = 12.0

    $b = \dfrac{12.0}{\tan 37°} = \dfrac{12.0}{0.7535\ldots} = 15.9245 \approx 15.9$

    $\sin 37° = \dfrac{opp}{hyp} = \dfrac{12.0}{c}$

    c(sin 37°) = 12.0

    $c = \dfrac{12.0}{\sin 37°} = \dfrac{12.0}{0.6018\ldots} = 19.939\ldots \approx 19.9$

13. c = 0.92, B = 49.9°

    A = 180° − 90° − 49.9° = 40.1°

    $\cos 49.9° = \dfrac{adj}{hyp} = \dfrac{a}{0.92}$

    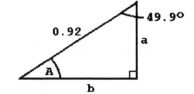

    a = 0.92(cos 49.9°)
      = 0.92(0.6441) = 0.59259 ≈ 0.59

    $\sin 49.0° = \dfrac{opp}{hyp} = \dfrac{b}{0.92}$

    b = 0.92(sin 49.9°) = 12.0 = 0.92(0.7649) = 0.70372... ≈ 0.70

Exercise 6.5

17. c = 54.40, A = 53.125°

B = 180° − 90° − 53.125° = 36.875°

$$\sin\ 53.125° = \frac{opp}{hyp} = \frac{a}{54.40}$$

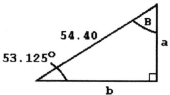

a = 54.40(sin 53.125°)
  = 54.40(0.79995) = 43.51709 ≈ 43.52

$$\cos\ 53.125° = \frac{adj}{hyp} = \frac{b}{54.40}$$

b = 54.40(cos 53.125°) = 54.40(0.60007) = 32.6439 ≈ 32.64

21. a = 15.0, c = 23.0

$$a^2 + b^2 = c^2$$

$$(15.0)^2 + b^2 = (23.0)^2$$

$$b^2 = (23.0)^2 - (15.0)^2$$

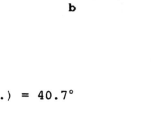

$$= 529 - 225 = 304$$

$$b = \sqrt{304} = 17.43559 ≈ 17.4$$

$$A = \sin^{-1}\left(\frac{opp}{hyp}\right) = \sin^{-1}\left(\frac{15.0}{23.0}\right) = \sin^{-1}(0.65217...) = 40.7°$$

B = 180° − 90° − 40.7° = 49.3°

25. b = 0.123, c = 0.456

$$a^2 + b^2 = c^2$$

$$a^2 + (0.123)^2 = (0.456)^2$$

$$a^2 = (0.456)^2 - (0.123)^2$$

$$= 0.207936 - 0.015129 = 0.192807$$

$$a = \sqrt{0.1928007} = 0.43909979 ≈ 0.439$$

$$B = \sin^{-1}\left(\frac{opp}{hyp}\right) = \sin^{-1}\left(\frac{0.123}{0.456}\right) = \sin^{-1}(0.269736) = 15.6°$$

A = 180° − 90° − 15.6° = 74.4°

29. tower = 20 feet,
     angle of depression = 7.6°

     θ = 180° − 90° − 7.6° = 82.4°

     $\tan 82.4° = \dfrac{pool}{tower} = \dfrac{pool}{20}$

        pool = 20(tan 82.4°) = 20(7.49465) = 149.893 ≈ 150 feet

33. bell tower = 48.5 feet,
     θ = angle of elevation

     a)   θ = 15.4°

           $\tan 15.4° = \dfrac{tower}{x} = \dfrac{48.5}{x}$

        x(tan 15.4°) = 48.5

           $x = \dfrac{48.5}{\tan 15.4°} = \dfrac{48.5}{0.27544} = 176.078 ≈ 176.1$ feet

     b)   θ = 61.2°

           $x = \dfrac{48.5}{\tan 61.2°} = \dfrac{48.5}{1.81899} = 26.663 ≈ 26.7$ feet

37. $\tan θ = \dfrac{opp}{adj}$, then opp = adj(tan θ)

     small triangle:
        opp = h, adj = x, θ = 61°

        h = x(tan 61°)

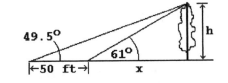

     large triangle:
        opp = h, adj = x + 50, θ = 49.5°

        h = (x + 50)(tan 49.5°)

                $h_{\text{small triangle}} = h_{\text{large triange}}$

                x(tan 61°) = (x + 50)(tan 49.5°)

                x(tan 61°) = x(tan 49.5°) + 50(tan 49.5°)

Exercise 6.5

37. Continued

$$x(\tan 61°) - x(\tan 49.5°) = 50(\tan 49.5°)$$

$$x(\tan 61° - \tan 49.5°) = 50(\tan 49.5°)$$

$$x = \frac{50(\tan 49.5°)}{\tan 61° - \tan 49.5°}$$

$$= \frac{50(1.17085)}{1.80405 - 1.17085} = 92.455$$

$$h = x(\tan 61°) = 92.455(1.80405) = 166.7936 \approx 167 \text{ feet}$$

41. adj = 900 feet, opp = h,
     angle of elevation = 58.24°

$$\tan 58.24° = \frac{\text{opp}}{\text{adj}} = \frac{h}{900}$$

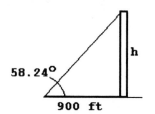

$$h = 900(\tan 58.24°)$$
$$= 900(1.61535)$$
$$= 1453.817 \approx 1454 \text{ feet}$$

Exercise 6.6

1. $x^2 + y^2 = 1^2$

$$(x - 0)^2 + (y - 0)^2 = 1^2$$

    Center is at $(0,0)$, $r = 1$

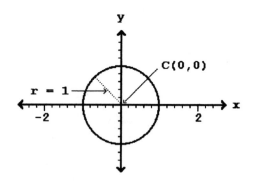

5. $x^2 + y^2 - 10x + 4y + 13 = 0$

$$(x^2 - 10x) + (y^2 + 4y) = -13$$

    To complete x: $\left(\frac{10}{2}\right)^2 = 5^2 = 25$,   To complete y: $\left(\frac{4}{2}\right)^2 = 2^2 = 4$

5.   Continued

$(x^2 - 10x + 25) + (y^2 + 4y + 4) = -13 + 25 + 4$

$(x - 5)^2 + (y + 2)^2 = 16$ Factor grouped terms

$(x - 5)^2 + (y - (-2))^2 = 4^2$ Rewrite to match formula
                                       with minus signs and r^2

Center is at $(5,-2)$, $r = 4$

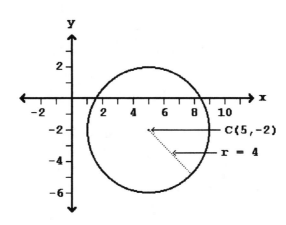

9. $y = x^2$, Vertex is at $(0,0)$

x	$y = x^2$
0	0
±1	1
±2	4
±3	9

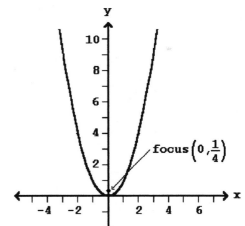

Find the focus:
   Use the point $Q(1,1)$
   which is on the parabola

$$4py = x^2$$
$$4p(1) = (1)^2$$
$$4p = 1$$

$$p = \frac{1}{4}, \quad \text{The focus is at } \left(0, \frac{1}{4}\right).$$

13. Draw a parabola with its vertex at the origin. Q(4.5,1.75) is a point on the parabola.

$$4py = x^2$$

$$4p(1.75) = (4.5)^2$$

$$7p = 20.25$$

$$p = 2.89286 \approx 2.9 \text{ ft}$$

Place the water container at the focus of the parabola which would be centered approximately 2.9 feet above the bottom of the reflector dish.

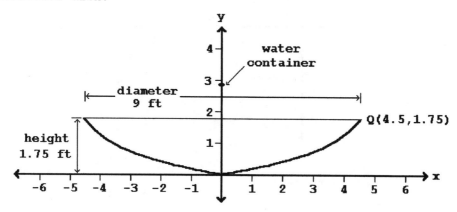

17. $4x^2 + 9y^2 = 36$

$$\frac{4x^2}{36} + \frac{9y^2}{36} = \frac{36}{36}$$

$$\frac{x^2}{9} + \frac{y^2}{4} = 1$$

<u>x-intercept (y = 0)</u>   <u>y-intercept (x = 0)</u>

$$\frac{x^2}{9} + \frac{(0)^2}{4} = 1 \qquad\qquad \frac{(0)^2}{9} + \frac{y^2}{4} = 1$$

$$\frac{x^2}{9} = 1 \qquad\qquad\qquad \frac{y^2}{4} = 1$$

$$x^2 = 9 \qquad\qquad\qquad y^2 = 4$$

$$x = \pm 3 \qquad\qquad\qquad y = \pm 2$$

17. Continued

    Finding the foci:

    $$a^2 = 9, \ b^2 = 4$$

    $$c^2 = a^2 - b^2$$

    $$= 9 - 4 = 5$$

    $$c = \pm\sqrt{5} \approx \pm 2.236$$

    Foci are located on

    the x-axis at $(-\sqrt{5}, 0)$

    and $(\sqrt{5}, 0)$.

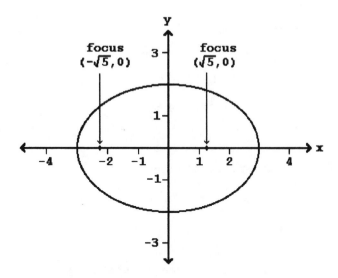

21. $16x^2 + 9y^2 = 144$

    $$\frac{16x^2}{144} + \frac{9y^2}{144} = \frac{144}{144}$$

    $$\frac{x^2}{9} + \frac{y^2}{16} = 1$$

    <u>x-intercept (y = 0)</u>   <u>y-intercept (x = 0)</u>

    $\dfrac{x^2}{9} + \dfrac{(0)^2}{16} = 1$ $\dfrac{(0)^2}{9} + \dfrac{y^2}{16} = 1$

    $\quad\quad\dfrac{x^2}{9} = 1$ $\quad\quad\dfrac{y^2}{16} = 1$

    $\quad\quad x^2 = 9$ $\quad\quad y^2 = 16$

    $\quad\quad x = \pm 3$ $\quad\quad y = \pm 4$

Exercise 6.6

21. Continued

    Finding the foci:

    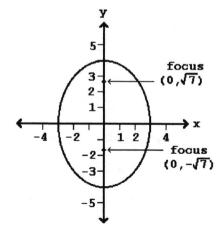

    $$a^2 = 9, \ b^2 = 16$$

    $$c^2 = b^2 - a^2$$

    $$= 16 - 9 = 7$$

    $$c = \pm\sqrt{7} \approx \pm 2.646$$

    Foci are located on

    the y-axis at $(0, -\sqrt{7})$

    and $(0, \sqrt{7})$.

25. Center an ellipse at the origin on a rectangular coordinate system so that the x-axis is 30 feet long and the y-axis is 24 feet long. This makes a = 15 and b = 12.

    $$\frac{x^2}{a^2} + \frac{y^2}{b^2} = \frac{x^2}{15^2} + \frac{y^2}{12^2} = \frac{x^2}{225} + \frac{y^2}{144} = 1$$

    Finding the foci:

    $$a^2 = 225, \ b^2 = 144$$

    $$c^2 = a^2 - b^2 = 225 - 144 = 81$$

    $$c = \pm\sqrt{81} = \pm 9$$

    Foci are located on the x-axis at $(-9, 0)$ and $(9, 0)$.

25. Continued

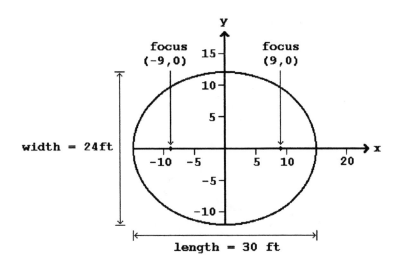

One person should stand 9 feet to the left from the center of the room along the longer axis which is the location of one focus. The other person should stand 9 feet to the right of the center at the other focus.

29. a) $x^2 - y^2 = 1$, $a^2 = 1$, $b^2 = 1$

<u>x-intercept (y = 0)</u>   <u>y-intercept (x = 0)</u>

$x^2 - (0)^2 = 1$ $(0)^2 - y^2 = 1$

$x^2 = 1$ $-y^2 = 1$

$x = \pm 1$ no solution; no y-intercepts

Locate a = ±1 on the x-axis and b = ±1 on the y-axis. Draw a rectangle with sides parallel to the axes going through these points. Now draw the diagonals of the rectangle. Then draw the branches of the hyperbola opening left and right.

Finding the foci:

$a^2 = 1$, $b^2 = 1$

$c^2 = a^2 + b^2 = 1 + 1 = 2$

$c = \pm\sqrt{2} \approx \pm 1.414$

Exercise 6.6

29.a) Continued

Foci are located on the x-axis at $(-\sqrt{2}, 0)$ and $(\sqrt{2}, 0)$.

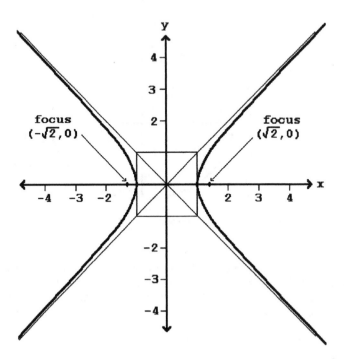

b) $y^2 - x^2 = 1$, $a^2 = 1$, $b^2 = 1$

<u>x-intercept (y = 0)</u>   <u>y-intercept (x = 0)</u>

$(0)^2 - x^2 = 1$ $y^2 - (0)^2 = 1$

$-x^2 = 1$ $y^2 = 1$

no solution; no x-intercepts $y = \pm 1$

Locate a = ±1 on the x-axis and b = ±1 on the y-axis. Draw
a rectangle with sides parallel to the axes going through
these points. Now draw the diagonals of the rectangle.
Then draw the branches of the hyperbola opening up and down.

29.b)   Continued

Finding the foci:

$a^2 = 1, \ b^2 = 1$

$c^2 = a^2 + b^2 = 1 + 1 = 2$

$c = \pm\sqrt{2} = \approx \pm 1.414$

Foci are located on the y-axis at $(0, -\sqrt{2})$ and $(0, \sqrt{2})$.

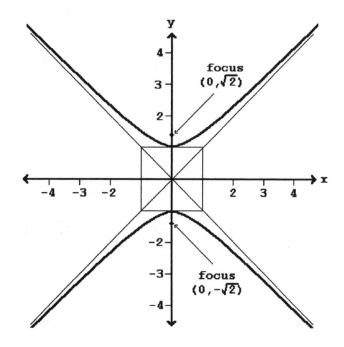

33. $25y^2 - 4x^2 = 100$

$$\frac{25y^2}{100} - \frac{4x^2}{100} = \frac{100}{100}$$

$$\frac{y^2}{4} - \frac{x^2}{25} = 1, \qquad a^2 = 25, \ b^2 = 4$$

33. Continued

x-intercept (y = 0)	y-intercept (x = 0)

$$\frac{(0)^2}{4} - \frac{x^2}{25} = 1 \qquad\qquad \frac{y^2}{4} - \frac{(0)^2}{25} = 1$$

$$-\frac{x^2}{25} = 1 \qquad\qquad\qquad \frac{y^2}{4} = 1$$

$$-x^2 = 25 \qquad\qquad\qquad y^2 = 4$$

no solution; no x-intercepts $\qquad$ y = ±2

Locate a = ±5 on the x-axis and b = ±2 on the y-axis. Draw a rectangle with sides parallel to the axes going through these points. Now draw the diagonals of the rectangle. Then draw the branches of the hyperbola opening up and down.

Finding the foci:

$$a^2 = 25, \ b^2 = 4$$

$$c^2 = a^2 + b^2 = 25 + 4 = 29$$

$$c = \pm\sqrt{29} \approx \pm 5.385$$

Foci are located on the y-axis at $(0, -\sqrt{29})$ and $(0, \sqrt{29})$.

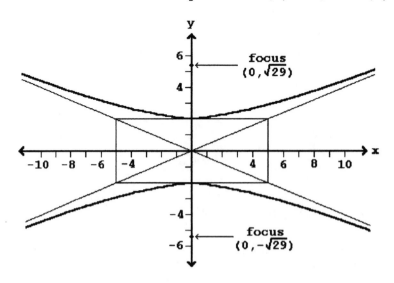

37. Omitted.

Exercise 6.7

1. Only one line parallel to the given line can be drawn through a point not on the given line.

5. A pair of distinct lines can intersect in only zero or one points.

9. A triangle can have zero to three right angles. (The sum of the three angles can be greater than 180°.) See the following drawings to visualize the different possible triangles.

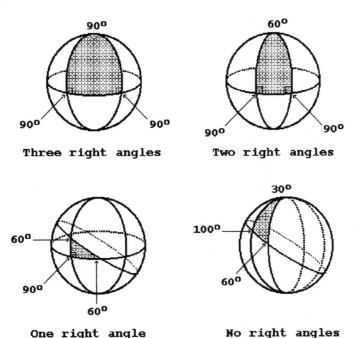

Three right angles   Two right angles

One right angle   No right angles

13. Infinitely many lines can be drawn parallel to the given line through a point not on the given line. See Figure 6.95 in your textbook.

17. A pair of distinct lines can meet in zero or one points in Poincaré's model of Lobachevskian geometry. See Figure 6.95 where AB and CD meet in zero points and CD and EF meet in one point.

21. Omitted.

25. Omitted.

Chapter 6 Review

1.   Omitted.

5.   a) Extending the line that is 6 yards long to
         the base creates a small square (s = 2 yd)
         and a trapezoid (b_1 = 6 + 2 = 8 yd, b_2 = 5
         yd and h = 12 - 2 = 10yd).

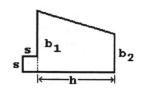

$$A = A_{square} + A_{trapezoid}$$

$$= s^2 + \frac{b_1 + b_2}{2}h = (2 \text{ yd})^2 + \frac{8 \text{ yd} + 5 \text{ yd}}{2}(10 \text{ yd})$$

$$= 4 \text{ yd}^2 + (13 \text{ yd})(5 \text{ yd}) = (4 + 65) \text{ yd}^2 = 69 \text{ yd}^2$$

   b) Draw a line across from the top of the 5
       yard side to the 6 yard side creating a
       triangle so that the length of the
       hypotenuse (the unmarked side) can be found.
       The base of the triangle = 10 yd, the height
       = (6 + 2 - 5) yd = 3 yd.

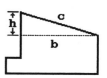

$$c^2 = a^2 + b^2 = 10^2 + 3^2 = 100 + 9 = 109, \ c = \sqrt{109}$$

To find the perimeter, start at one corner and add the
lengths.

$$P = (5 + 12 + 2 + 2 + 6 + \sqrt{109}) \text{ yd}$$

$$= (27 + \sqrt{109}) \text{ yd} = (27 + 10.4403) \text{ yd}$$

$$= 37.4403 \text{ yd} \approx 37.4 \text{ yd}$$

9. Rectangular box: L = 14 in., w = 8 in., h = 10 in.
    Small cube: s = 4 in.

    $V = V_{\text{rectangular box}} - V_{\text{small cube}}$

    $= L \cdot w \cdot h - s^3$

    $= (14 \text{ in.})(8 \text{ in.})(10 \text{ in.}) - (4 \text{ in.})^3$

    $= (1120 - 64) \text{ in.}^3 = 1056 \text{ in.}^3$

    To find the surface area, note that if the missing cube's sides
    are "popped out" it completes the rectangular box.

    $A = 2(A_{\text{bottom}}) + 2(A_{\text{side}}) + 2(A_{\text{front}})$

    $= 2(L \cdot w) + 2(w \cdot h) + 2(L \cdot h)$

    $= 2(14 \text{ in.})(8 \text{ in.}) + 2(8 \text{ in.})(10 \text{ in.}) + 2(14 \text{ in.})(10 \text{ in.})$

    $= 224 \text{ in.}^2 + 160 \text{ in.}^2 + 280 \text{ in.}^2 = 664 \text{ in.}^2$

13. L = (18 - 2(2)) in. = (18 - 4) in. = 14 in.
    w = (12 - 2(2)) in. = (12 - 4) in. = 8 in.
    h = 2 in.

    a) $V = L \cdot w \cdot h$

          $= (14 \text{ in.})(8 \text{ in.})(2 \text{ in.}) = 224 \text{ in.}^3$

    b) $A = A_{\text{bottom}} + 2(A_{\text{front}}) + 2(A_{\text{side}})$

          $= L \cdot w + 2(L \cdot h) + 2(w \cdot h)$

          $= (14 \text{ in.})(8 \text{ in.}) + 2(14 \text{ in.})(2 \text{ in.}) + 2(8 \text{ in.})(2 \text{ in.})$

          $= 112 \text{ in.}^2 + 56 \text{ in.}^2 + 32 \text{ in.}^2 = 200 \text{ in.}^2$

17. Sphere: r = 1 cubit

    a)   Egyptian approximation of pi $\left(\pi = \dfrac{256}{81}\right)$

          $V = \dfrac{4}{3}\pi r^3 = \dfrac{4}{3} \cdot \dfrac{256}{81}(1 \text{ cubit})^3 = \dfrac{1024}{243}(1 \text{ cubit})^3$

          $= 4.21399177 \text{ cubits}^3 \approx 4.2140 \text{ cubits}^3$

17. Continued

    b) Calculator value of pi.

    $$A = \frac{4}{3}\pi r^3 = \frac{4}{3}\pi(1 \text{ cubit})^3 = \frac{4}{3}\pi \text{ cubits}^3$$

    $$= 4.188790205 \text{ cubits}^3 \approx 4.1888 \text{ cubits}^3$$

    c) Error of Egyptian calculation relative to the calculator value.

    $$A_{Egyptian} - A_{Calculator} = 4.21399177 - 4.188790205$$

    $$= 0.025201564 \text{ cubits}^3$$

    $$\text{relative error} = \frac{0.025201564 \text{ cubits}^3}{4.188790205 \text{ cubits}^3} = 0.006016 \approx 0.6\%$$

21. Rectangular Box:
    L = 5 ft,
    w = 3.5 ft,
    h = 2.5 ft

    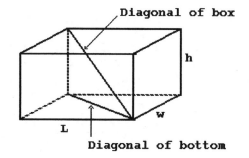

    Diagonal of box

    Diagonal of bottom

    c = diagonal on bottom of box

    d = diagonal of the box

    $$c^2 = L^2 + w^2$$

    $$c^2 = (5 \text{ ft})^2 + (3.5 \text{ ft})^2$$

    $$= 25 \text{ ft}^2 + 12.25 \text{ ft}^2 = 37.25 \text{ ft}^2$$

    $$d^2 = c^2 + h^2 = 37.25 \text{ ft}^2 + (2.5 \text{ ft})^2$$

    $$= 37.25 \text{ ft}^2 + 6.25 \text{ ft}^2 = 43.5 \text{ ft}^2$$

    $$d = \sqrt{43.5 \text{ ft}^2} = 6.59545 \text{ ft} \approx 6.5 \text{ ft}$$

The length of the longest object that will fit inside the box is 6.5 feet. Note: It is appropriate in this situation to not round up so that the object will fit inside the box.

25. Given: $\angle CBA = \angle DAB$

$BC = AD$

Prove: $\angle CAD = \angle DBC$

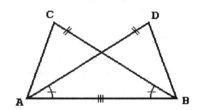

Statements	Reasons
1. $\angle CBA = \angle DAB$	1. Given
2. $BC = AD$	2. Given
3. $AB = AB$	3. Anything equals itself
4. $\triangle ACB \cong \triangle BDA$	4. SAS
5. $\angle CAB = \angle DBA$	5. Corresponding parts of congruent triangles are equal
6. $\angle CAD = \angle DBC$	6. Equals subtracted from equals are equal

29. a) $9x^2 - 4y^2 = 36$

$$\frac{9x^2}{36} - \frac{4y^2}{36} = \frac{36}{36}$$

$$\frac{x^2}{4} - \frac{y^2}{9} = 1, \quad a^2 = 4, \quad b^2 = 9$$

<u>x-intercept (y = 0)</u>   <u>y-intercept (x = 0)</u>

$$\frac{x^2}{4} - \frac{(0)^2}{9} = 1 \qquad\qquad \frac{(0)^2}{4} - \frac{y^2}{9} = 1$$

$$\frac{x^2}{4} = 1 \qquad\qquad\qquad -\frac{y^2}{9} = 1$$

$$x^2 = 4 \qquad\qquad\qquad -y^2 = 9$$

$$x = \pm 2 \qquad\quad \text{no solution; no x-intercepts}$$

29.a) Continued

Locate a = ±2 on the x-axis and b = ±3 on the y-axis. Draw a rectangle with sides parallel to the axes going through these points. Now draw the diagonals of the rectangle. Then draw the branches of the hyperbola opening left and right.

Finding the foci:

$$a^2 = 4, \ b^2 = 9$$

$$c^2 = a^2 + b^2 = 4 + 9 = 13$$

$$c = \pm\sqrt{13} \approx \pm 3.606$$

Foci are located on the x-axis at $(-\sqrt{13}, 0)$ and $(\sqrt{13}, 0)$.

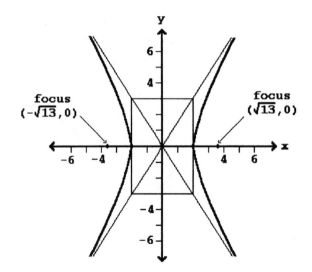

b) $4y^2 - 9x^2 = 36$

$$\frac{4y^2}{36} - \frac{9x^2}{36} = \frac{36}{36}$$

$$\frac{y^2}{9} - \frac{x^2}{4} = 1, \ a^2 = 4, \ b^2 = 9$$

29.b) Continued

<u>x-intercept (y = 0)</u>   <u>y-intercept (x = 0)</u>

$$\frac{(0)^2}{9} - \frac{x^2}{4} = 1 \qquad\qquad \frac{y^2}{9} - \frac{(0)^2}{4} = 1$$

$$-\frac{x^2}{4} = 1 \qquad\qquad\qquad \frac{y^2}{9} = 1$$

$$-x^2 = 4 \qquad\qquad\qquad y^2 = 9$$

no solution; no x-intercepts   y = ±3

Locate a = ±2 on the x-axis and b = ±3 on the y-axis. Draw a rectangle with sides parallel to the axes going through these points. Now draw the diagonals of the rectangle. Then draw the branches of the hyperbola opening up and down.

Finding the foci:

$$a^2 = 4, \ b^2 = 9$$

$$c^2 = a^2 + b^2 = 4 + 9 = 13$$

$$c = \pm\sqrt{13} \approx \pm3.606$$

Foci are located on the y-axis at $(0, -\sqrt{13})$ and $(0, \sqrt{13})$.

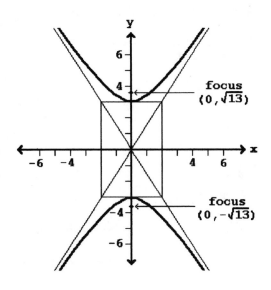

33.　　B = 180° − 90° − 25.4° = 64.6°

$$\sin 25.4° = \frac{opp}{hyp} = \frac{a}{56.1}$$

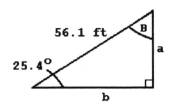

a = 56.1(sin 25.4°)
　 = 56.1(0.42894) = 24.063 ≈ 24.1 ft

$$\cos 25.4° = \frac{adj}{hyp} = \frac{b}{56.1}$$

b = 56.1(cos 25.4°) = 56.1(0.90334) = 50.677 ≈ 50.7 ft

37.　a)　　The Parallel Postulate is:　Given a line and a point not on
　　　　　that line, there is one and only one line through the point
　　　　　parallel to the original line.

　　b)　　Alternative 1:　Given a line and a point not on the line,
　　　　　there are no lines through the point parallel to the
　　　　　original line (parallels do not exist).

　　　　　Alternative 2:　Given a line and a point not on the line,
　　　　　there are at least two lines through the point parallel to
　　　　　the original line.

　　c)　　Alternative 1:　Riemannian geometry uses the sphere as a
　　　　　model where lines are defined as great circles that
　　　　　encompass the sphere.

　　　　　Alternative 2:　Lobachevskian geometry uses the interior of
　　　　　a circle as a model and defines lines as either the diameter
　　　　　of the disk or a circular arc connecting two points on the
　　　　　boundary of the disk.

　　d)

Alternative 1

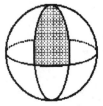

Riemannian Geometry

Alternative 2

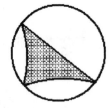

Lobachevskian Geometry

7 Matrices and Markov Chains

Exercise 7.0

1. a) 3 rows and 2 columns, so the dimensions of matrix A are 3 × 2.
   b) Matrix A is neither a row matrix, a column matrix, nor a square matrix.

5. a) 1 row and 2 columns, so the dimensions of matrix E are 1 × 2.
   b) Matrix E is a row matrix.

9. a) 3 rows and 1 column, so the dimensions of matrix J are 3 × 1.
   b) Matrix J is a column matrix.

13. c_{21} = 41   row 2, column 1 of matrix C (Exercise 3)

17. g_{12} = -11   row 1, column 2 of matrix G (Exercise 7)

21. a)
$$AC = \begin{bmatrix} 5 & 0 \\ 22 & -3 \\ 18 & 9 \end{bmatrix} \begin{bmatrix} 23 \\ 41 \end{bmatrix} = \begin{bmatrix} 5 \cdot 23 + 0 \cdot 41 \\ 22 \cdot 23 + -3 \cdot 41 \\ 18 \cdot 23 + 9 \cdot 41 \end{bmatrix} = \begin{bmatrix} 115 \\ 383 \\ 783 \end{bmatrix}$$

   b)
$$CA = \begin{bmatrix} 23 \\ 41 \end{bmatrix} \begin{bmatrix} 5 & 0 \\ 22 & -3 \\ 8 & 9 \end{bmatrix} = \text{does not exist}$$

   2 × 1   3 × 2
       ↑ ↑
   different, so we can't multiply

25. a)
$$CG = \begin{bmatrix} 23 \\ 41 \end{bmatrix} \begin{bmatrix} 12 & -11 & 5 \\ -9 & 4 & 0 \\ 1 & 9 & 5 \end{bmatrix} = \text{does not exist}$$

   2 × 1   3 × 3
       ↑ ↑
   different, so we can't multiply

Exercise 7.0

25. Continued

b)
$$GC = \begin{bmatrix} 12 & -11 & 5 \\ -9 & 4 & 0 \\ 1 & 9 & 5 \end{bmatrix} \begin{bmatrix} 23 \\ 41 \end{bmatrix} = \text{does not exist}$$

$$\quad 3 \times 3 \qquad 2 \times 1$$
$$\qquad\quad \uparrow \qquad\quad \uparrow$$

different, so we can't multiply

29. a)
$$AF = \begin{bmatrix} 5 & 0 \\ 22 & -3 \\ 18 & 9 \end{bmatrix} \begin{bmatrix} -2 & 10 \\ 4 & -3 \end{bmatrix} = \begin{bmatrix} 5 \cdot -2 + 0 \cdot 4 & 5 \cdot 10 + 0 \cdot -3 \\ 22 \cdot -2 + -3 \cdot 4 & 22 \cdot 10 + -3 \cdot -3 \\ 18 \cdot -2 + 9 \cdot 4 & 18 \cdot 10 + 9 \cdot -3 \end{bmatrix}$$

$$= \begin{bmatrix} -10 & 50 \\ -56 & 229 \\ 0 & 153 \end{bmatrix}$$

b)
$$FA = \begin{bmatrix} -2 & 10 \\ 4 & -3 \end{bmatrix} \begin{bmatrix} 5 & 0 \\ 22 & -3 \\ 18 & 9 \end{bmatrix} = \text{does not exist}$$

$$\quad 2 \times 2 \quad 3 \times 2$$
$$\qquad \uparrow \quad \uparrow$$

different, so we can't multiply

33. Table could be rewritten as the matrix T and the purchase as matrix P. Find PT, not TP, so that the dimensions match.

$$T = \begin{bmatrix} 1.25 & 1.30 \\ 0.95 & 1.10 \end{bmatrix} \qquad P = [2 \quad 1]$$

$$PT = [2 \quad 1] \begin{bmatrix} 1.25 & 1.30 \\ 0.95 & 1.10 \end{bmatrix} = [2 \cdot 1.25 + 1 \cdot 0.95 \quad 2 \cdot 1.30 + 1 \cdot 1.10]$$

$$= [3.45 \quad 3.70]$$

The price at Blondie's would be $3.45 and the price at SliceMan's would be $3.70.

37.
$$BC = \begin{bmatrix} -3 & 0 & -1 \\ 1 & 4 & -2 \end{bmatrix} \begin{bmatrix} 0 \\ 5 \\ -1 \end{bmatrix} = \begin{bmatrix} -3 \cdot 0 + 0 \cdot 5 + -1 \cdot -1 \\ 1 \cdot 0 + 4 \cdot 5 + -2 \cdot -1 \end{bmatrix} = \begin{bmatrix} 1 \\ 22 \end{bmatrix}$$

$$A(BC) = \begin{bmatrix} -4 & 5 \\ 2 & 3 \end{bmatrix} \begin{bmatrix} 1 \\ 22 \end{bmatrix} = \begin{bmatrix} -4 \cdot 1 + 5 \cdot 22 \\ 2 \cdot 1 + 3 \cdot 22 \end{bmatrix} = \begin{bmatrix} 106 \\ 68 \end{bmatrix}$$

37. Continued

$$AB = \begin{bmatrix} -4 & 5 \\ 2 & 3 \end{bmatrix} \begin{bmatrix} -3 & 0 & -1 \\ 1 & 4 & -2 \end{bmatrix}$$

$$= \begin{bmatrix} -4 \cdot -3 + 5 \cdot 1 & -4 \cdot 0 + 5 \cdot 4 & -4 \cdot -1 + 5 \cdot -2 \\ 2 \cdot -3 + 3 \cdot 1 & 2 \cdot 0 + 3 \cdot 4 & 2 \cdot -1 + 3 \cdot -2 \end{bmatrix} = \begin{bmatrix} 17 & 20 & -6 \\ -3 & 12 & -8 \end{bmatrix}$$

$$(AB)C = \begin{bmatrix} 17 & 20 & -6 \\ -3 & 12 & -8 \end{bmatrix} \begin{bmatrix} 0 \\ 5 \\ -1 \end{bmatrix} = \begin{bmatrix} 17 \cdot 0 + 20 \cdot 5 + -6 \cdot -1 \\ -3 \cdot 0 + 12 \cdot 5 + -8 \cdot -1 \end{bmatrix} = \begin{bmatrix} 106 \\ 68 \end{bmatrix}$$

41.

$$GC = \begin{bmatrix} 1 & 0 & 0 \\ 0 & 1 & 0 \\ 0 & 0 & 1 \end{bmatrix} \begin{bmatrix} 4 & 1 & -1 \\ 5 & 12 & 3 \end{bmatrix} = \text{does not exist}$$

$$\begin{array}{cc} 3 \times 3 & 2 \times 1 \\ \uparrow & \uparrow \end{array}$$

different, so we can't multiply

45. Omitted.

49.

$$BA = \begin{bmatrix} 7 & 9 \\ -8 & 0 \end{bmatrix} \begin{bmatrix} 5 & -7 \\ 3 & 9 \end{bmatrix} = \begin{bmatrix} 7 \cdot 5 + 9 \cdot 3 & 7 \cdot -7 + 9 \cdot 9 \\ -8 \cdot 5 + 0 \cdot 3 & -8 \cdot -7 + 0 \cdot 9 \end{bmatrix} = \begin{bmatrix} 62 & 32 \\ -40 & 56 \end{bmatrix}$$

53. Answers for both a) and b) are the same because matrix multiplication is associative.

$$(BF)C = B(FC) = \begin{bmatrix} -23840 & 1688 & -7720 \\ 109348 & 23813 & -16663 \end{bmatrix}$$

Exercise 7.1

1. a) Given:
        p(current purchase is KickKola) = 0.14
      Complement:
        p(current purchase is not KickKola) = 1 - 0.14 = 0.86

   b)   K   K′
      P = [0.14   0.86]

Exercise 7.1

5. a)   Given:
               p(Silver's Gym) = 0.48
               p(Fitness Lab) = 0.37
         Complement:
               p(ThinNFit) = 1 - (0.48 + 0.37) = 0.15

    b)   S   F   T
         P = [0.48 0.37 0.15]

9. a)   Given:
               p(will buy home|currently rents) = 0.12
               p(will sell and then rent|currently owns) = 0.03

         Complements:
               p(will not buy home|currently rents) = 1 - 0.12 = 0.88
               p(will not sell home|currently owns) = 1 - 0.03 = 0.97

    b)   H   H'
         $T = \begin{bmatrix} 0.97 & 0.03 \\ 0.12 & 0.88 \end{bmatrix}$ H
                                                 H'

13. Refer to Exercises 1 and 7.

            K   K'
     P = [0.14 0.86]

            K   K'
     $T = \begin{bmatrix} 0.63 & 0.37 \\ 0.12 & 0.88 \end{bmatrix}$ K
                                             K'

    a) PT^1 = [0.14 0.86] $\begin{bmatrix} 0.63 & 0.37 \\ 0.12 & 0.88 \end{bmatrix}$

               = [0.14•0.63 + 0.86•0.12   0.14•0.37 + 0.86•0.88]
               = [0.1914   0.8086]

         KickKola's market share at the next following purchase will
         be 19% if current trends continue.

    b) PT^2 = (PT)T = [0.1914 0.8086] $\begin{bmatrix} 0.63 & 0.37 \\ 0.12 & 0.88 \end{bmatrix}$

               = [0.1914•0.63 + 0.8086•0.12
                                   0.1914•0.37 + 0.8086•0.88]
               = [0.217614   0.782386]

         KickKola's market share at the second following purchase
         will be 22% if current trends continue.

17.

$$P = \begin{matrix} S & S' \\ [0.41 & 0.59] \end{matrix}$$

$$T = \begin{matrix} & S & S' \\ & \begin{bmatrix} 0.12 & 0.88 \\ 0.31 & 0.69 \end{bmatrix} & \begin{matrix} S \\ S' \end{matrix} \end{matrix}$$

Find the second following purchase. $\left(\dfrac{4 \text{ years}}{2}\right) = 2$

$$PT^1 = [0.41 \quad 0.59] \begin{bmatrix} 0.12 & 0.88 \\ 0.31 & 0.69 \end{bmatrix}$$

$$= [0.2321 \quad 0.7679]$$

$$PT^2 = (PT)T = [0.2321 \quad 0.7679] \begin{bmatrix} 0.12 & 0.88 \\ 0.31 & 0.69 \end{bmatrix}$$

$$= [0.265901 \quad 0.734099]$$

Sierra Cruiser's market share in four years at the second following purchase will be 27%.

21. a) Cropland = 413 million acres
Potential cropland = 127 million acres
Noncropland = 856 million acres
Total cropland = 413 + 127 + 856 = 1396 million acres

$$P = \begin{matrix} C & PC & NC \\ \begin{bmatrix} \dfrac{413}{1396} & \dfrac{127}{1396} & \dfrac{856}{1396} \end{bmatrix} \end{matrix} = \begin{matrix} C & PC & NC \\ [0.296 & 0.091 & 0.613] \end{matrix}$$

b)

	New Use of Land (1977)			
Previous Use of Land (1967)	cropland	potential cropland	noncropland	total 1967
cropland	413 – 34 = 379	17	35	431
potential cropland	34	127 – 17 = 110	0	144
noncropland	0	0	856 – 35 = 821	821
total 1977	413	127	856	1396

Exercise 7.1

21.b) Continued

To calculate probabilities we must compare the new data (1977) to the previous data (1967).

$$
T = \begin{bmatrix} \frac{379}{431} & \frac{17}{431} & \frac{35}{431} \\[2mm] \frac{34}{144} & \frac{110}{144} & \frac{0}{144} \\[2mm] \frac{0}{821} & \frac{0}{821} & \frac{821}{821} \end{bmatrix}
\begin{matrix} C & PC & NC \end{matrix}
= \begin{bmatrix} 0.879 & 0.040 & 0.081 \\ 0.236 & 0.764 & 0 \\ 0 & 0 & 1 \end{bmatrix}
$$

With column labels C, PC, NC above each matrix.

c) Find the next following state.

$$
PT = [0.296 \quad 0.091 \quad 0.613] \begin{bmatrix} 0.879 & 0.040 & 0.081 \\ 0.236 & 0.764 & 0 \\ 0 & 0 & 1 \end{bmatrix}
$$

$$
= [0.281660 \quad 0.081364 \quad 0.636976]
$$

$$
= [0.282 \quad 0.081 \quad 0.637]
$$

In 1987, the amount of land of each type in acres will be:
cropland = 0.282(1396) = 394 million
potential cropland = 0.081(1396) = 113 million
noncropland = 0.637(1396) = 889 million

d) Find the second following state.

$$
(PT)T = [0.282 \quad 0.081 \quad 0.637] \begin{bmatrix} 0.879 & 0.040 & 0.081 \\ 0.236 & 0.764 & 0 \\ 0 & 0 & 1 \end{bmatrix}
$$

$$
= [0.266994 \quad 0.073164 \quad 0.659842]
$$

$$
= [0.267 \quad 0.073 \quad 0.660]
$$

In 1997, the amount of land of each type in acres will be:
cropland = 0.267(1396) = 373 million
potential cropland = 0.073(1396) = 102 million
noncropland = 0.660(1396) = 921 million

25. Omitted.

29. Omitted.

Exercise 7.2

1. Substitute (4,1) into the system of equations:
    $$3x - 5y = 3(4) - 5(1) = 12 - 5 = 7$$
    $$2x + 2y = 2(4) + 2(1) = 8 + 2 = 10$$
    Both equations are satisfied, so (4,1) is a solution.

5. Substitute (4,-1,2) into the system of equations:
    $$2x + 3y - z = 2(4) + 3(-1) - (2) = 8 - 3 - 2 = 3$$
    $$x + y + z = (4) + (-1) + (2) = 4 - 1 + 2 = 5$$
    $$10x - 2y = 10(4) - 2(-1) = 40 + 2 = 42 \quad \text{not} \quad 3$$
    The third equation is not satisfied, so (4,-1,2) is not a solution for the system.

9. a) $4x + 3y = 12$ $\quad\quad\quad\quad\quad$ $8x + 6y = 24$
    $\quad\quad\quad$ $3y = -4x + 12$ $\quad\quad\quad$ $6y = -8x + 24$

    $\quad\quad\quad$ $y = -\dfrac{4}{3}x + 4$ $\quad\quad\quad$ $y = -\dfrac{8}{6}x + \dfrac{24}{6} = -\dfrac{4}{3}x + 4$

    The slope of each equation is $-\dfrac{4}{3}$ and the y-intercept is 4.

    b)   Therefore, the equations are the same and there are an infinite number of solutions.

13. This system could have a unique solution since it has as many unique equations as it has unknowns.

17. This system will not have a single solution, it will have either no solution or an infinite number of solutions since it has fewer equations (2) than unknowns (3).

21. $3x - 7y = 27$ $\quad$ multiply by 4 $\quad$ $12x - 28y = 108$
    $4x - 5y = 23$ $\quad$ multiply by -3 $\quad$ $\underline{-12x + 15y = -69}$
    $\quad\quad\quad\quad\quad\quad$ add together $\quad\quad$ $0x - 13y = \quad 39$

    $\quad\quad\quad\quad\quad\quad$ solve for y $\quad\quad\quad\quad$ $y = \dfrac{39}{-13} = -3$

    Find x by substitution $\quad$ $3x - 7(-3) = 27$
    $\quad\quad\quad\quad\quad\quad\quad\quad\quad\quad$ $3x + 21 = 27$
    $\quad\quad\quad\quad\quad\quad\quad\quad\quad\quad\quad$ $3x = 6$
    $\quad\quad\quad\quad\quad\quad\quad\quad\quad\quad\quad\quad$ $x = 2$

    Solution is (2,-3)

    Check solution (2,-3): $\quad$ $3x - 7y = 3(2) - 7(-3) = 27$
    $\quad\quad\quad\quad\quad\quad\quad\quad\quad\quad$ $4x - 5y = 4(2) - 5(-3) = 23$

25. 2x + 7y = 11   All variables are on the left and all
     3x – 2y = 15   coefficients are showing.

$$\begin{bmatrix} 2 & 7 & 11 \\ 3 & -2 & 15 \end{bmatrix}$$

29. 2x + 3y – 7z = 53   All variables are on the left and
     5x – 2y + 12z = 19   missing coefficients have been
     1x + 1y + 1z = 55   added.

$$\begin{bmatrix} 2 & 3 & -7 & 53 \\ 5 & -2 & 12 & 19 \\ 1 & 1 & 1 & 55 \end{bmatrix}$$

33. $$\begin{bmatrix} 2 & 6 & 10 \\ -2 & 1 & 4 \end{bmatrix}$$

R1 ÷ 2:R1 $$\begin{bmatrix} 1 & 3 & 5 \\ -2 & 1 & 4 \end{bmatrix}$$

2R1 + R2:R2 $$\begin{bmatrix} 1 & 3 & 5 \\ 0 & 7 & 14 \end{bmatrix}$$

37. $$\begin{bmatrix} 2 & 2 & 4 & 12 \\ 2 & -1 & 4 & 3 \\ 1 & 2 & -9 & 2 \end{bmatrix}$$

R1 ÷ 2:R1 $$\begin{bmatrix} 1 & 1 & 2 & 6 \\ 2 & -1 & 4 & 3 \\ 1 & 2 & -9 & 2 \end{bmatrix}$$

–2R1 + R2:R2 $$\begin{bmatrix} 1 & 1 & 2 & 6 \\ 0 & -3 & 0 & -9 \\ 1 & 2 & -9 & 2 \end{bmatrix}$$

–1R1 + R3:R3 $$\begin{bmatrix} 1 & 1 & 2 & 6 \\ 0 & -3 & 0 & -9 \\ 0 & 1 & -11 & -4 \end{bmatrix}$$

41. x = 0.5281   Read down the column, take a right turn
     y = –0.6205   at 1, and read the number at the end of
     z = 0   the row.
       or (0.5281,–0.6205,0)

45.   Step 1: Write with all variables to the left and all coefficients
                  showing.

$$2x + 4y = 12$$
$$3x - 1y = 4$$

Step 2: Rewrite in matrix form

$$\begin{bmatrix} 2 & 4 & 12 \\ 3 & -1 & 4 \end{bmatrix}$$

Step 3: Row operations

First column:

$$R1 \div 2:R1 \quad \begin{bmatrix} 1 & 2 & 6 \\ 3 & -1 & 4 \end{bmatrix}$$

$$-3R1 + R2:R2 \quad \begin{bmatrix} 1 & 2 & 6 \\ 0 & -7 & -14 \end{bmatrix}$$

Second column:

$$R2 \div -7:R2 \quad \begin{bmatrix} 1 & 2 & 6 \\ 0 & 1 & 2 \end{bmatrix}$$

$$-2R2 + R1:R1 \quad \begin{bmatrix} 1 & 0 & 2 \\ 0 & 1 & 2 \end{bmatrix}$$

Step 4: Read the solution. $x = 2$, $y = 2$ or $(2,2)$

Step 5: *Check the solution (2,2):*

$$2x + 4y = 2(2) + 4(2) = 12$$
$$3x - y = 3(2) - (2) = 4$$

49.   Step 1: Write with all variables to the left and all coefficients showing.

$$5x + 1y - 1z = 17$$
$$2x + 5y + 2z = 0$$
$$3x + 1y + 1z = 11$$

Step 2: Rewrite in matrix form

$$\begin{bmatrix} 5 & 1 & -1 & 17 \\ 2 & 5 & 2 & 0 \\ 3 & 1 & 1 & 11 \end{bmatrix}$$

Exercise 7.2

49. Continued

    Step 3: Row operations

        First column:

$$R1 \div 5{:}R1 \quad \begin{bmatrix} 1 & 0.2 & -0.2 & 3.4 \\ 2 & 5 & 2 & 0 \\ 3 & 1 & 1 & 11 \end{bmatrix}$$

$$\begin{array}{l} \\ -2R1 + R2{:}R2 \\ -3R1 + R3{:}R3 \end{array} \begin{bmatrix} 1 & 0.2 & -0.2 & 3.4 \\ 0 & 4.6 & 2.4 & -6.8 \\ 0 & 0.4 & 1.6 & 0.8 \end{bmatrix}$$

        Second column:

$$R2 \div 4.6{:}R2 \quad \begin{bmatrix} 1 & 0.2 & -0.2 & 3.4 \\ 0 & 1 & 0.5217 & -1.4783 \\ 0 & 0.4 & 1.6 & 0.8 \end{bmatrix}$$

$$\begin{array}{l} -0.2R2 + R1{:}R1 \\ \\ -0.4R2 + R3{:}R3 \end{array} \begin{bmatrix} 1 & 0 & -0.3043 & 3.6957 \\ 0 & 1 & 0.5217 & -1.4783 \\ 0 & 0 & 1.3913 & 1.3913 \end{bmatrix}$$

        Third column:

$$R3 \div 1.3913{:}R3 \quad \begin{bmatrix} 1 & 0 & -0.3043 & 3.6957 \\ 0 & 1 & 0.5217 & -1.4783 \\ 0 & 0 & 1 & 1 \end{bmatrix}$$

$$\begin{array}{l} 0.3043R3 + R1{:}R1 \\ -0.5217R3 + R2{:}R2 \end{array} \begin{bmatrix} 1 & 0 & 0 & 4 \\ 0 & 1 & 0 & -2 \\ 0 & 0 & 1 & 1 \end{bmatrix}$$

    Step 4: Read the solution. $x = 4$, $y = -2$, $z = 1$ or $(4,-2,1)$

    Step 5: *Check the solution* $(4,-2,1)$:

        $5x + y - z = 5(4) + (-2) - (1) = 17$
        $2x + 5y + 2z = 2(4) + 5(-2) + 2(1) = 0$
        $3x + y + z = 3(4) + (-2) + (1) = 11$

53. Step 1: Write with all variables to the left and all coefficients showing. Missing coefficients have been added.

        $1x - 1y + 4z = -13$
        $2x + 0y - 1z = 12$
        $3x + 1y + 0z = 25$

53. Continued

Step 2: Rewrite in matrix form

$$\begin{bmatrix} 1 & -1 & 4 & -13 \\ 2 & 0 & -1 & 12 \\ 3 & 1 & 0 & 25 \end{bmatrix}$$

Step 3: Row operations

First column:

$$\begin{array}{c} \\ -2R1 + R2:R2 \\ -3R1 + R3:R3 \end{array} \begin{bmatrix} 1 & -1 & 4 & -13 \\ 0 & 2 & -9 & 38 \\ 0 & 4 & -12 & 64 \end{bmatrix}$$

Second column:

$$R2 \div 2:R2 \begin{bmatrix} 1 & -1 & 4 & -13 \\ 0 & 1 & -4.5 & 19 \\ 0 & 4 & -12 & 64 \end{bmatrix}$$

$$\begin{array}{c} R2 + R1:R1 \\ \\ -4R2 + R3:R3 \end{array} \begin{bmatrix} 1 & 0 & -0.5 & 6 \\ 0 & 1 & -4.5 & 19 \\ 0 & 0 & 6 & -12 \end{bmatrix}$$

Third column:

$$R3 \div 6:R3 \begin{bmatrix} 1 & 0 & -0.5 & 6 \\ 0 & 1 & -4.5 & 19 \\ 0 & 0 & 1 & -2 \end{bmatrix}$$

$$\begin{array}{c} 0.5R3 + R1:R1 \\ 4.5R3 + R2:R2 \end{array} \begin{bmatrix} 1 & 0 & 0 & 5 \\ 0 & 1 & 0 & 10 \\ 0 & 0 & 1 & -2 \end{bmatrix}$$

Step 4: Read the solution. $x = 5$, $y = 10$, $z = -2$ or $(5, 10, -2)$

Step 5: *Check the solution (5,10,-2):*

$x - y + 4z = (5) - (10) + 4(-2) = -13$
$2x - z = 2(5) - (-2) = 12$
$3x + y = 3(5) + (10) = 25$

Exercise 7.2

57. The first equation was ignored: $3x - 5y + z = 12$

   Step 1: Write with all variables to the left and all coefficients showing.

   $$2x + 1y + 1z = 3$$
   $$5x - 4y + 1z = 0$$
   $$1x + 1y + 1z = 4$$

   Step 2: Rewrite in matrix form

   $$\begin{bmatrix} 2 & 1 & 1 & 3 \\ 5 & -4 & 1 & 0 \\ 1 & 1 & 1 & 4 \end{bmatrix}$$

   Step 3: Row operations

   First column:

   $$\begin{matrix} \text{Move R3:R1} \\ \text{Move R1:R2} \\ \text{Move R2:R3} \end{matrix} \begin{bmatrix} 1 & 1 & 1 & 4 \\ 2 & 1 & 1 & 3 \\ 5 & -4 & 1 & 0 \end{bmatrix}$$

   $$\begin{matrix} \\ -2R1 + R2:R2 \\ -5R1 + R3:R3 \end{matrix} \begin{bmatrix} 1 & 1 & 1 & 4 \\ 0 & -1 & -1 & -5 \\ 0 & -9 & -4 & -20 \end{bmatrix}$$

   Second column:

   $$\begin{matrix} \\ -R2:R2 \\ \\ \end{matrix} \begin{bmatrix} 1 & 1 & 1 & 4 \\ 0 & 1 & 1 & 5 \\ 0 & -9 & -4 & -20 \end{bmatrix}$$

   $$\begin{matrix} -R2 + R1:R1 \\ \\ 9R2 + R3:R3 \end{matrix} \begin{bmatrix} 1 & 0 & 0 & -1 \\ 0 & 1 & 1 & 5 \\ 0 & 0 & 5 & 25 \end{bmatrix}$$

   Third column:

   $$\begin{matrix} \\ \\ R3 \div 5:R3 \end{matrix} \begin{bmatrix} 1 & 0 & 0 & -1 \\ 0 & 1 & 1 & 5 \\ 0 & 0 & 1 & 5 \end{bmatrix}$$

   $$\begin{matrix} \\ -1R3 + R2:R2 \\ \\ \end{matrix} \begin{bmatrix} 1 & 0 & 0 & -1 \\ 0 & 1 & 0 & 0 \\ 0 & 0 & 1 & 5 \end{bmatrix}$$

57. Continued

Step 4: Read the solution. $x = -1$, $y = 0$, $z = 5$ or $(-1,0,5)$

Step 5: *Check the solution* $(-1,0,5)$:

$$2x + y + z = 2(-1) + (0) + (5) = 3$$
$$5x - 4y + z = 5(-1) - 4(0) + (5) = 0$$
$$x + y + z = (-1) + (0) + (5) = 4$$

Step 6: *Check the solution in the first equation.*

$$3x - 5y + z = 3(-1) - 5(0) + (5) = -3 + 5 = 2 \text{ not } 12$$

Hence this system has no solutions.

61. Step 1: Write with all variables to the left and all coefficients showing.

$$1x + 1y + 1z = 3$$
$$2x - 1y + 1z = 2$$

Step 2: Rewrite in matrix form

$$\begin{bmatrix} 1 & 1 & 1 & 3 \\ 2 & -1 & 1 & 2 \end{bmatrix}$$

Step 3: Row operations

First column:

$$-2R1 + R2:R2 \quad \begin{bmatrix} 1 & 1 & 1 & 3 \\ 0 & -3 & -1 & -4 \end{bmatrix}$$

Second column:

$$R2 \div -3:R2 \quad \begin{bmatrix} 1 & 1 & 1 & 3 \\ 0 & 1 & 1/3 & 4/3 \end{bmatrix}$$

$$-R2 + R1:R1 \quad \begin{bmatrix} 1 & 0 & 2/3 & 5/3 \\ 0 & 1 & 1/3 & 4/3 \end{bmatrix}$$

Step 4: Write the new system of equations

$$1x + 0y + \frac{2}{3}z = \frac{5}{3} \quad \text{or} \quad 3x + 2z = 5$$

$$0x + 1y + \frac{1}{3}z = \frac{4}{3} \quad \text{or} \quad 3y + z = 4$$

61. Continued

Step 5: Let z equal other numbers and find the related x and y.

z = 0   3x + 2z = 5   3y + z = 4
              3x + 2(0) = 5   3y + (0) = 4
                  3x = 5   3y = 4
                   x = 5/3   y = 4/3

z = 1   3x + 2z = 5   3y + z = 4
              3x + 2(1) = 5   3y + (1) = 4
                  3x = 5 - 2   3y = 4 - 1
                  3x = 3   3y = 3
                   x = 1   y = 1

z = -1   3x + 2z = 5   3y + z = 4
              3x + 2(-1) = 5   3y + (-1) = 4
                  3x = 5 + 2   3y = 4 + 1
                  3x = 7   3y = 5
                   x = 7/3   y = 5/3

Some selected solutions would be:
   (5/3,4/3,0), (1,1,1), (7/3,5/3,-1)

65. (1)   5x + 1y - 1z = 17
     (2)   2x + 5y + 2z = 0
     (3)   3x + 1y + 1z = 11

Eliminate z from equation (1) and equation (2):

(1)   10x + 2y - 2z = 34   2 times equation (1)
(2)   2x + 5y + 2z = 0   no change
(4)   12x + 7y + 0z = 34   sum = equation (4)

Eliminate z from equation (1) and equation (3):

(1)   5x + 1y - 1z = 17   no change
(3)   3x + 1y + 1z = 11   no change
(5)   8x + 2y + 0z = 28   sum = equation (5)

Solve equation (4) and equation (5) by elimination:

12x + 7y = 34   multiply by 2   24x + 14y = 68
 8x + 2y = 28   multiply by -3   -24x - 6y = -84
                 add together   0x + 8y = -16

                 solve for y $y = \dfrac{-16}{8} = -2$

65. Continued

    Find x by substitution   8x + 2(-2) = 28
                                  8x - 4 = 28
                                     8x = 32
                                      x = 4

    Substitute x = 4 and y = -2 into equation (1):

    $$5x + y - z = 17$$
    $$5(4) + (-2) - z = 17$$
    $$20 - 2 - z = 17$$
    $$18 - z = 17$$
    $$-z = -1$$
    $$z = 1$$

    Solution is (4,-2,1)

    Check solution (4,-2,1) in equation (2) and equation (3):

    (2)   2x + 5y + 2z = 2(4) + 5(-2) + 2(1) = 8 -10 + 2 = 0
    (3)   3x + 1y + 1z = 3(4) + (-2) + (1) = 12 -2 + 1 = 11

69. Omitted.

73. Write the system of equations in standard form:
       1x + 1y + 1z = 2.35
       -2x + 3y - 3z = 6.2
       1.2x - 0.4y + 2.1z = -2.64

    Rewrite the system in matrix form:

    $$\begin{bmatrix} 1.0 & 1.0 & 1.0 & 2.35 \\ -2.0 & 3.0 & -3.0 & 6.20 \\ 1.2 & -0.4 & 2.1 & -2.64 \end{bmatrix}$$

    Matrix is 3 × 4

    Row Operations:
         Column 1: No Division
                   Add   +2•Row1 + Row2:Row2
                   Add   -1.2•Row1 + Row3:Row3

         Column 2: Divide Row2 by 5 or Multiply by 1/5
                   Add   -1•Row2 + Row1:Row1
                   Add   +1.6•Row2 + Row3:Row3

73. Continued

> Column 3: Divide Row3 by 0.58 or Multiply by 1/0.58
> Add $-1.2 \cdot$Row3 + Row1:Row1
> Add $+0.2 \cdot$Row3 + Row2:Row2

Solution: $x = 4.25$, $y = 1.5$, $z = -3.4$ or $(4.25, 1.5, -3.4)$

Check the solution by substituting into the original equations:
$x + y + z = (4.25) + (1.5) + (-3.4) = 2.35$
$-2y + 3z - 3x = -2(4.25) + 3(1.5) - 3(-3.4) = 6.2$
$1.2x - 0.4y + 2.1z = 1.2(4.25) - 0.4(1.5) + 2.1(-3.4)$
$= -2.64$

77. Write the system of equations in standard form:
$1x - 2y + 1z = 0$
$-15x + 3y + 2z = 0$
$9x + 2y + 7z = 29$

Rewrite the system in matrix form:

$$\begin{bmatrix} 1 & -2 & 1 & 0 \\ -15 & 3 & 2 & 0 \\ 9 & 2 & 7 & 29 \end{bmatrix}$$

Matrix is 3 × 4

Row Operations:
> Column 1: No Division
> Add $+15 \cdot$Row1 + Row2:Row2
> Add $-9 \cdot$Row1 + Row3:Row3

> Column 2: Divide Row2 by -27 or Multiply by $-1/27$
> Add $+2 \cdot$Row2 + Row1:Row1
> Add $-20 \cdot$Row2 + Row3:Row3

> Column 3: Divide Row3 by 10.5925 or Multiply by 1/10.5925
> Add $+0.2592 \cdot$Row3 + Row1:Row1
> Add $+0.6296 \cdot$Row3 + Row2:Row2

Solution: $x = 0.70979$, $y = 1.72377$, $z = 2.73776$
or $(0.70979, 1.72377, 2.73776)$

Check the solution by substituting into the original equations.

81. The system of equations is in standard form

    Rewrite the system in matrix form:

    $$\begin{bmatrix} 3.9 & 4.9 & -3.9 & -4.2 & 4.2 \\ 4.7 & 3.7 & 3.8 & 9.7 & 1.1 \\ 7.0 & -7.2 & 3.9 & 3.6 & 5.0 \\ 8.5 & 6.3 & -9.0 & -3.7 & 0.0 \end{bmatrix}$$

    Matrix dimensions are 4×5

    Row Operations:
        Column 1: Divide Row1 by 3.9 or Multiply by 1/3.9
                Add   -4.7•Row1 + Row2:Row2
                Add   -7.0•Row1 + Row3:Row3
                Add   -8.5•Row1 + Row4:Row4

        Column 2: Divide Row2 by -2.2051 or Multiply by -1/2.2051
                Add   -1.2564•Row2 + Row1:Row1
                Add   +15.994•Row2 + Row3:Row3
                Add   +4.3794•Row2 + Row4:Row4

        Column 3: Divide Row3 by -50.754 or Multiply by -1/50.754
                Add   -3.8430•Row3 + Row1:Row1
                Add   +3.8546•Row3 + Row2:Row2
                Add   +17.381•Row3 + Row4:Row4

        Column 4: Divide Row4 by 8.99029 or Multiply by 1/8.99029
                Add   -0.0698•Row4 + Row1:Row1
                Add   -0.5917•Row4 + Row2:Row2
                Add   -1.8901•Row4 + Row3:Row3

    Solution: $x_1 = 0.88295$, $x_2 = 0.48206$, $x_3 = 1.64039$, $x_4 = -1.1409$
           or $(0.88295, 0.48206, 1.64039, -1.1409)$

    Check the solution by substituting into the original equations.

Exercise 7.2

85. The system of equations is in standard form

Rewrite the system in matrix form:

$$\begin{bmatrix} 6.3 & 5.9 & -3.9 & -4.7 & 9.1 & 71.8 \\ 6.4 & 9.2 & 5.1 & 2.2 & -7.6 & 81.1 \\ 3.4 & -7.0 & 2.9 & 3.5 & 4.2 & 15.0 \\ 5.7 & 3.6 & -9.0 & -2.4 & 3.3 & 100.0 \\ 3.6 & 4.3 & -5.7 & -6.4 & -2.8 & 53.0 \end{bmatrix}$$

Matrix dimensions are 5 × 6

Row Operations:
Column 1: Divide Row1 by 6.3 or Multiply by 1/6.3
          Add −6.4•Row1 + Row2:Row2
          Add −3.4•Row1 + Row3:Row3
          Add −5.7•Row1 + Row4:Row4
          Add −3.6•Row1 + Row5:Row5

Column 2: Divide Row2 by 3.20634 or Multiply by 1/3.20634
          Add −0.9365•Row2 + Row1:Row1
          Add +10.184•Row2 + Row3:Row3
          Add +1.7380•Row2 + Row4:Row4
          Add −0.9286•Row2 + Row5:Row5

Column 3: Divide Row3 by 33.7875 or Multiply by 1/33.7875
          Add +3.2658•Row3 + Row1:Row1
          Add −2.8262•Row3 + Row2:Row2
          Add +0.5591•Row3 + Row4:Row4
          Add +6.0957•Row3 + Row5:Row5

Column 4: Divide Row4 by 6.09968 or Multiply by 1/6.09968
          Add +0.0584•Row4 + Row1:Row1
          Add +0.1827•Row4 + Row2:Row2
          Add −0.8343•Row4 + Row3:Row3
          Add +0.6483•Row4 + Row5:Row5

Column 5: Divide Row5 by −14.492 or Multiply by −1/14.492
          Add −0.9809•Row5 + Row1:Row1
          Add +1.1668•Row5 + Row2:Row2
          Add −0.4419•Row5 + Row3:Row3
          Add +2.4528•Row5 + Row4:Row4

Solution: $x_1 = 10.5623$, $x_2 = 2.40364$, $x_3 = -4.9053$,
        $x_4 = 4.07244$, $x_5 = -0.9796$
        or $(10.5623, 2.40364, -4.9053, 4.07244, -0.9796)$

Check the solution by substituting into the original equations.

Exercise 7.3

1. Step 1: Create the transition matrix T.

    $$T = \begin{bmatrix} 0.1 & 0.9 \\ 0.2 & 0.8 \end{bmatrix}$$

    Step 2: Create the equilibrium matrix L.

    L = [x   y]

    Step 3: Find and simplify the system of equations described by LT = L.

    $$[x \quad y] \begin{bmatrix} 0.1 & 0.9 \\ 0.2 & 0.8 \end{bmatrix} = [x \quad y]$$

    [0.1x + 0.2y   0.9x + 0.8y] = [x   y]

    Yields: 0.1x + 0.2y = x
              0.9x + 0.8y = y

    Combining like terms: -0.9x + 0.2y = 0
                             0.9x - 0.2y = 0

    Step 4: Discard any redundant equations and include x + y = 1.

    The above equations are the same so use only once.

    The system to be solved is:
        x + y = 1
        0.9x - 0.2y = 0

    Step 5: Solve the resulting system.

    $$
    \begin{array}{lll}
    x + y = 1 & \text{times 0.2} & 0.2x + 0.2y = 0.2 \\
    0.9x - 0.2\,y = 0 & & \underline{0.9x - 0.2y = 0} \\
    & \text{add together} & 1.1x \qquad\quad = 0.2
    \end{array}
    $$

    solve for x $x = \dfrac{0.2}{1.1} = \dfrac{2}{11} = 0.1818$

    Find y by substitution:
        0.1818 + y = 1
                y = 1 - 0.1818 = 0.8182

    Resulting matrix is L = [0.1818   0.8182] ≈ [0.18   0.82]

Exercise 7.3

1. Continued

    Step 6: *Check the work by verifying that LT = L.*

    $$LT = [0.1818 \quad 0.8182] \begin{bmatrix} 0.1 & 0.9 \\ 0.2 & 0.8 \end{bmatrix}$$

    $$= [0.18182 \quad 0.81818] \quad \approx \quad [0.1818 \quad 0.8182]$$

5. Refer to Exercise 13 in Section 7.3.

    Step 1: Create the transition matrix T.

    $$T = \begin{bmatrix} 0.63 & 0.37 \\ 0.12 & 0.88 \end{bmatrix}$$

    Step 2: Create the equilibrium matrix L.

    $$L = [x \quad y]$$

    Step 3: Find and simplify the system of equations described by LT = L.

    $$[x \quad y] \begin{bmatrix} 0.63 & 0.37 \\ 0.12 & 0.88 \end{bmatrix} = [x \quad y]$$

    $$[0.63x + 0.12y \quad 0.37x + 0.88y] = [x \quad y]$$

    Yields: $0.63x + 0.12y = x$
    $0.37x + 0.88y = y$

    Combining like terms: $-0.37x + 0.12y = 0$
    $0.37x - 0.12y = 0$

    Step 4: Discard any redundant equations and include
    $x + y = 1$.

    The above equations are the same so use only once.

    The system to be solved is:
    $x + y = 1$
    $0.37x - 0.12y = 0$

    Step 5: Solve the resulting system.

$x + y = 1$	times 0.12	$0.12x + 0.12y = 0.12$
$0.37x - 0.12y = 0$		$\underline{0.37x - 0.12y = 0}$
	add together	$0.49x \qquad = 0.12$
	solve for x	$x = \dfrac{0.12}{0.49} = 0.2449$

240

5. Continued

Find y by substitution:
        0.2449 + y = 1
               y = 1 - 0.2449 = 0.7551

    Resulting matrix is L = [0.2449 0.7551]

Step 6: *Check the work by verifying that LT = L.*

$$LT = [0.2449 \quad 0.7551] \begin{bmatrix} 0.63 & 0.37 \\ 0.12 & 0.88 \end{bmatrix}$$

        = [0.244899 0.755101] ≈ [0.2449 0.7551] = L

KickKola's long-range market share will be 24%.

9. Refer to Exercise 9 in Section 7.1.

Step 1: Create the transition matrix T.

$$T = \begin{bmatrix} 0.97 & 0.03 \\ 0.12 & 0.88 \end{bmatrix}$$

Step 2: Create the equilibrium matrix L.

    L = [x y]

Step 3: Find and simplify the system of equations described by LT = L.

$$[x \quad y] \begin{bmatrix} 0.97 & 0.03 \\ 0.12 & 0.88 \end{bmatrix} = [x \quad y]$$

    [0.97x + 0.12y   0.03x + 0.88y] = [x   y]

Yields: 0.97x + 0.12y = x
          0.03x + 0.88y = y

Combining like terms: -0.03x + 0.12y = 0
                         0.03x - 0.12y = 0

Step 4: Discard any redundant equations and include x + y = 1. The above equations are the same so use only once.

    The system to be solved is:
        x + y = 1
        0.03x - 0.12y = 0

Exercise 7.3

9. Continued

Step 5: Solve the resulting system.

$$x + y = 1 \qquad \text{times } 0.12 \qquad 0.12x + 0.12y = 0.12$$
$$0.03x - 0.12y = 0 \qquad\qquad\qquad\qquad \underline{0.03x - 0.12y = 0}$$
$$\text{add together} \qquad 0.15x \qquad\qquad = 0.12$$

$$\text{solve for } x \qquad x = \frac{0.12}{0.15} = 0.8$$

Find y by substitution:
$$0.8 + y = 1$$
$$y = 1 - 0.8 = 0.2$$

Resulting matrix is $\qquad L = [0.8 \quad 0.2]$

Step 6: *Check the work by verifying that LT = L.*

$$LT = [0.8 \quad 0.2] \begin{bmatrix} 0.97 & 0.03 \\ 0.12 & 0.88 \end{bmatrix}$$

$$= [0.8 \quad 0.2] \quad = \quad L$$

This means that 80% of Metropolis's residents will eventually own their own homes and 20% of Metropolis's residents will rent. The assumptions that are made in this prediction are that current housing trends will continue as expressed in the transition matrix and that the resident's moving plans are realized.

13. Step 1: Create the transition matrix T.

$$T = \begin{bmatrix} 0.60 & 0.10 & 0.12 & 0.08 & 0.07 & 0.03 \\ 0.09 & 0.64 & 0.07 & 0.08 & 0.09 & 0.03 \\ 0.19 & 0.10 & 0.54 & 0.09 & 0.07 & 0.01 \\ 0.05 & 0.06 & 0.05 & 0.69 & 0.12 & 0.03 \\ 0.03 & 0.05 & 0.13 & 0.16 & 0.61 & 0.02 \\ 0.18 & 0.15 & 0.18 & 0.16 & 0.17 & 0.16 \end{bmatrix} \begin{matrix} T \\ N \\ R \\ J \\ D \\ H \end{matrix}$$

with column headers T N R J D H

Step 2: Create the equilibrium matrix L.

Let x_1 = Toyonda, x_2 = Nissota, x_3 = Reo
x_4 = Henry J, x_5 = DeSota, x_6 = Hugo

$$L = [x_1 \quad x_2 \quad x_3 \quad x_4 \quad x_5 \quad x_6]$$

13. Continued

Step 3: Find and simplify the system of equations described by LT = L.

Multiply matrices as shown in Exercises above to obtain the following equations:

$0.60x_1 + 0.09x_2 + 0.19x_3 + 0.05x_4 + 0.03x_5 + 0.18x_6 = x_1$
$0.10x_1 + 0.64x_2 + 0.10x_3 + 0.06x_4 + 0.05x_5 + 0.15x_6 = x_2$
$0.12x_1 + 0.07x_2 + 0.54x_3 + 0.05x_4 + 0.13x_5 + 0.18x_6 = x_3$
$0.08x_1 + 0.08x_2 + 0.09x_3 + 0.69x_4 + 0.16x_5 + 0.16x_6 = x_4$
$0.07x_1 + 0.09x_2 + 0.07x_3 + 0.12x_4 + 0.61x_5 + 0.17x_6 = x_5$
$0.03x_1 + 0.03x_2 + 0.01x_3 + 0.03x_4 + 0.02x_5 + 0.16x_6 = x_6$

Combining like terms:

$-0.40x_1 + 0.09x_2 + 0.19x_3 + 0.05x_4 + 0.03x_5 + 0.18x_6 = 0$
$0.10x_1 - 0.36x_2 + 0.10x_3 + 0.06x_4 + 0.05x_5 + 0.15x_6 = 0$
$0.12x_1 + 0.07x_2 - 0.46x_3 + 0.05x_4 + 0.13x_5 + 0.18x_6 = 0$
$0.08x_1 + 0.08x_2 + 0.09x_3 - 0.31x_4 + 0.16x_5 + 0.16x_6 = 0$
$0.07x_1 + 0.09x_2 + 0.07x_3 + 0.12x_4 - 0.39x_5 + 0.17x_6 = 0$
$0.03x_1 + 0.03x_2 + 0.01x_3 + 0.03x_4 + 0.02x_5 - 0.84x_6 = 0$

Step 4: Discard any redundant equations and include

$x_1 + x_2 + x_3 + x_4 + x_5 + x_6 = 1.$

If the first five equations above are added together they will equal the last equation. Therefore the last equation is discarded.

Step 5: Solve the resulting system. Use the computer software or your calculator.

Rewrite the system in matrix form:

$$\begin{bmatrix} 1.0 & 1.0 & 1.0 & 1.0 & 1.0 & 1.0 & 1 \\ -0.4 & 0.09 & 0.19 & 0.05 & 0.03 & 0.18 & 0 \\ 0.10 & -0.36 & 0.10 & 0.06 & 0.05 & 0.15 & 0 \\ 0.12 & 0.07 & -0.46 & 0.05 & 0.13 & 0.18 & 0 \\ 0.08 & 0.08 & 0.09 & -0.31 & 0.16 & 0.16 & 0 \\ 0.07 & 0.09 & 0.07 & 0.12 & -0.39 & 0.17 & 0 \end{bmatrix}$$

Matrix dimensions are 6 × 7

Exercise 7.3

13. Continued

    Row Operations:
       Column 1: No Division
                Add +0.40•Row1 + Row2:Row2
                Add −0.10•Row1 + Row3:Row3
                Add −0.12•Row1 + Row4:Row4
                Add −0.08•Row1 + Row5:Row5
                Add −0.07•Row1 + Row6:Row6

       Column 2: Divide Row2 by 0.49 or Multiply by 1/0.49
                Add −1.00•Row2 + Row1:Row1
                Add +0.46•Row2 + Row3:Row3
                Add +0.05•Row2 + Row4:Row4
                Add −0.02•Row2 + Row6:Row6

       Column 3: Divide Row3 by 0.55387
                             or Multiply by 1/0.55387
                Add +0.2040•Row3 + Row1:Row1
                Add −1.2040•Row3 + Row2:Row2
                Add +0.5197•Row3 + Row4:Row4
                Add −0.0100•Row3 + Row5:Row5
                Add +0.0240•Row3 + Row6:Row6

       Column 4: Divide Row4 by 0.33483
                             or Multiply by 1/0.33483
                Add −0.2226•Row4 + Row1:Row1
                Add −0.0870•Row4 + Row2:Row2
                Add −0.6905•Row4 + Row3:Row3
                Add +0.3970•Row4 + Row5:Row5
                Add −0.0483•Row4 + Row6:Row6

       Column 5: Divide Row5 by 0.53092
                             or Multiply by 1/0.53092
                Add +0.0036•Row5 + Row1:Row1
                Add −0.0085•Row5 + Row2:Row2
                Add +0.1570•Row5 + Row3:Row3
                Add −1.1522•Row5 + Row4:Row4
                Add +0.5177•Row5 + Row6:Row6

       Column 6: Divide Row6 by −0.85488
                           or Multiply by −1/0.85488
                Add +0.4086•Row6 + Row1:Row1
                Add +0.2985•Row6 + Row2:Row2
                Add +0.0650•Row6 + Row3:Row3
                Add −0.1301•Row6 + Row4:Row4
                Add −1.6422•Row6 + Row5:Row5

13. Continued

> Solution: $x_1 = 0.17845$, $x_2 = 0.17724$, $x_3 = 0.16712$,
> $x_4 = 0.25499$, $x_5 = 0.19377$, $x_6 = 0.02841$
> or L = [0.178 0.177 0.167 0.255 0.194 0.028]

Step 6: *Check the work by verifying that LT = L.*

> First verify:
> $x_1 + x_2 + x_3 + x_4 + x_5 + x_6$
> $= 0.17845 + 0.17724 + 0.16712 + 0.25499 + 0.19377$
> $+ 0.02841 = 0.99998 \approx 1$

> Verify all the equations from Step 3: (Only the first one is done as an example.)

> $0.60x_1 + 0.09x_2 + 0.19x_3 + 0.05x_4 + 0.03x_5 + 0.18x_6 = x_1$

> $= 0.60(0.17845) + 0.09(0.17724) + 0.19(0.16712)$
> $+ 0.05(0.25499) + 0.03(0.19377) + 0.18(0.02841)$
> $= 0.1784508 \approx 0.17845 = x_1$

The Moldavian market share for each brand of automobile is as follows:

x_1 = Toyonda = 17.8%
x_2 = Nissota = 17.7%
x_3 = Reo = 16.7%
x_4 = Henry J = 25.5%
x_5 = DeSota = 19.4%
x_6 = Hugo = 2.8%

Note: Due to round-off error the percentages given total only 99.9%.

Chapter 7 Review

1. The given matrix is neither a row matrix, a column matrix, nor a square matrix. The dimensions are 3 × 2.

5. $\begin{bmatrix} 1 & 6 \\ 8 & 4 \end{bmatrix} \begin{bmatrix} -3 & 6 \\ 0 & -5 \end{bmatrix} = \begin{bmatrix} 1 \cdot -3 + 6 \cdot 0 & 1 \cdot 6 + 6 \cdot -5 \\ 8 \cdot -3 + 4 \cdot 0 & 8 \cdot 6 + 4 \cdot -5 \end{bmatrix} = \begin{bmatrix} -3 & -24 \\ -24 & 28 \end{bmatrix}$

9. The product does not exist because the dimensions do not match (3 × 4 and 2 × 3).

13. $5x - 7y = -29$ multiply by 2 $10x - 14y = -58$
    $2x + 4y = 2$ multiply by -5 $\underline{-10x - 20y = -10}$
                    add together $0x - 34y = -68$

solve for y $y = \dfrac{-68}{-34} = 2$

Find x by substitution $5x - 7(2) = -29$
                            $5x - 14 = -29$
                               $5x = -15$
                                  $x = -3$

The solution is $(-3,2)$

Check $(-3,2)$: $5x - 7y = 5(-3) - 7(2) = -29$
                  $2x + 4y = 2(-3) + 4(2) = 2$

17.   T   T'
    $P = [0.12 \quad 0.88]$

        T   T'
$T = \begin{bmatrix} 0.62 & 0.38 \\ 0.16 & 0.84 \end{bmatrix} \begin{matrix} T \\ T' \end{matrix}$

a)   Find the first following purchase

$PT^1 = [0.12 \quad 0.88] \begin{bmatrix} 0.62 & 0.38 \\ 0.16 & 0.84 \end{bmatrix}$

       $= [0.2152 \quad 0.7848]$

After three years, Toyonda Motors' market share will be 21.5%.

b)   Find the second following purchase. $\left(\dfrac{6 \text{ years}}{3}\right) = 2$

$PT^2 = (PT)T = [0.2152 \quad 0.7848] \begin{bmatrix} 0.62 & 0.38 \\ 0.16 & 0.84 \end{bmatrix}$

       $= [0.258992 \quad 0.741008]$

Toyonda Motors' market share in six years (second following purchase) will be 25.9%

c)   Step 1: Create the transition matrix T.

$T = \begin{bmatrix} 0.62 & 0.38 \\ 0.16 & 0.84 \end{bmatrix}$

Step 2: Create the equilibrium matrix L.

$L = [x \quad y]$

17.c) Continued

Step 3:   Find and simplify the system of equations
described by LT = L.

$$[x \quad y] \begin{bmatrix} 0.62 & 0.38 \\ 0.16 & 0.84 \end{bmatrix} = [x \quad y]$$

$$[0.62x + 0.16y \quad 0.38x + 0.84y] = [x \quad y]$$

Yields: 0.62x + 0.16y = x
          0.38x + 0.84y = y

Combining like terms: −0.38x + 0.16y = 0
                         0.38x − 0.16y = 0

Step 4:   Discard any redundant equations and include
x + y = 1.

The above equations are the same so use only once.

The system to be solved is:
x + y = 1
0.38x − 0.16y = 0

Step 5:   Solve the resulting system.

x + y = 1   times 0.16   0.16x + 0.16y = 0.16
0.38x − 0.16y = 0   0.38x − 0.16y = 0
                   add together 0.54x   = 0.16

solve for x $x = \dfrac{0.16}{0.54} = 0.2963$

Find y by substitution:
0.2963 + y = 1
        y = 1 − 0.2963 = 0.7037

Resulting matrix is L = [0.2963   0.7037]

Step 6:   *Check the work by verifying that LT = L.*

$$LT = [0.2963 \quad 0.7037] \begin{bmatrix} 0.62 & 0.38 \\ 0.16 & 0.84 \end{bmatrix}$$

$$= [0.296298 \quad 0.703702] \approx [0.2963 \quad 0.7037] = L$$

Toyonda Motors' long-range market share will be 29.6%.

d)   We always assume that the current trends reflected in the
transition matrix will be continued into the future and the
survey accurately reflects future purchases.

8   Linear Programming

<u>**Exercise 8.0**</u>

1.   **Solve the inequality for y:** $3x + y < 4 \;\rightarrow\; y < -3x + 4$

Graph the line: $y = -3x + 4$ is a dashed line (= is not part of the inequality) with slope -3 and y-intercept 4. Start at $(0,4)$ on the y-axis and from that point rise -3 (move three units down) and run 1 (move one unit to the right).

$$\text{slope} = m = \frac{\text{rise}}{\text{run}} = -3 = \frac{-3}{1} = \frac{\text{three units down}}{\text{one unit right}}$$

Shade in one side of the line: The graph of the inequality is the set of all points below the line (values of y decrease if we move down).

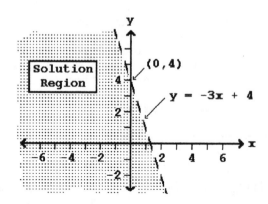

5.   **Solve the inequality for y:**   There is no y in $x \geq 4$.

Graph the line: $x = 4$ is a solid line (= is a part of the inequality) through any x-coordinate of 4 such as $(4,0)$, $(4,2)$ and $(4,3)$. This makes a vertical line.

Shade in one side of the line: The graph of the inequality is the set of all points to the right of the line (values of x increase if we move to the right).

5.   Continued

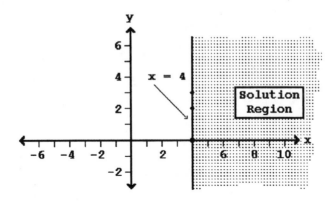

9.   a)

Original Inequality	Slope-Intercept Form	Associated Equation	Graph of the Inequality
y > 2x + 1	y > 2x + 1	y = 2x + 1	all points above the line with slope 2 and y-intercept 1
y ≤ -x + 4	y ≤ -x + 4	y = -x + 4	all points on or below the line with slope -1 and y-intercept 4

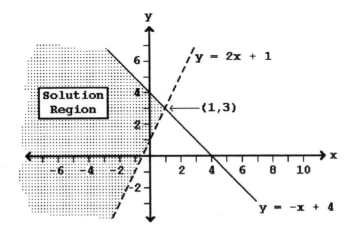

b)   The region of solutions is unbounded.

Exercise 8.0

9. Continued

   c) There is one corner point where $y = 2x + 1$ intersects $y = -x + 4$. Use the elimination method.

$$
\begin{array}{ll}
y = 2x + 1 & \rightarrow \quad y = 2x + 1 \\
y = -x + 4 & \rightarrow \quad \underline{-y = x - 4} \\
& 0 = 3x - 3
\end{array}
$$

$$
-3x = -3 \quad \rightarrow \quad x = 1
$$

$$
y = 2x + 1 = 2(1) + 1 = 3
$$

   Thus, the corner point is $(1, 3)$.

13. a)

Original Inequality	Slope-Intercept Form	Associated Equation	Graph of the Inequality
$x + 2y \leq 4$	$2y \leq -x + 4 \rightarrow$ $y \leq -\frac{1}{2}x + 2$	$y = -\frac{1}{2}x + 2$	all points on or below the line with slope $-(1/2)$ and y-intercept 2
$3x - 2y \leq -12$	$-2y \leq -3x - 12$ $y \geq \frac{3}{2}x + 6$	$y = \frac{3}{2}x + 6$	all points on or above the line with slope $(3/2)$ and y-intercept 6
$x - y < -7$	$-y < -x - 7 \rightarrow$ $y > x + 7$	$y = x + 7$	all points above the line with slope 1 and y-intercept 7

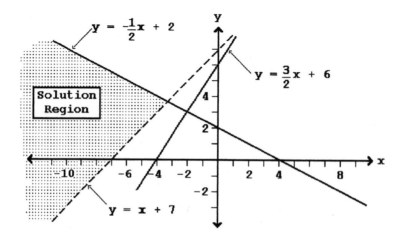

250

13. **Continued**

   b)   The region of solutions is unbounded.

   c)   There is one corner point where $x + 2y = 4$ intersects $x - y = -7$. Use the elimination method.

$$
\begin{array}{llll}
x + 2y = 4 & \rightarrow & x + 2y = 4 \\
x - y = -7 & \rightarrow & \underline{-x + y = 7} \\
& & 3y = 11
\end{array}
$$

$$3y = 11 \quad \rightarrow \quad y = \frac{11}{3} = 3\frac{2}{3}$$

$$x - y = -7 \quad \rightarrow \quad x - \frac{11}{3} = -7 \quad \rightarrow \quad x = -\frac{21}{3} + \frac{11}{3} = -\frac{10}{3} = -3\frac{1}{3}$$

   Thus, the corner point is $\left(-3\frac{1}{3},\ 3\frac{2}{3}\right)$.

17. a)

Original Inequality	Associated Equation	Graph of the Inequality
Slope-Intercept Form		
$15x + 22y \leq 510$ $22y \leq -15x + 510 \rightarrow$ $y \leq -\frac{15}{22}x + \frac{510}{22}$	$y = -\frac{15}{22}x + \frac{510}{22}$	all points on or below the line with slope $-(15/22)$ and y-intercept $510/22$
$35x + 12y \leq 600$ $12y \leq -35x + 600 \rightarrow$ $y \leq -\frac{35}{12}x + 50$	$y = -\frac{35}{12}x + 50$	all points on or below the line with slope $-(35/12)$ and y-intercept 50
$x + y > 10$ $y > -x + 10$	$y = -x + 10$	all points above the line with slope -1 and y-intercept 10
$x \geq 0$ (not applicable)	$x = 0$	all points on or to the right of the y-axis
$y \geq 0$ $y \geq 0$	$y = 0$	all points on or above the x-axis

17.a)   Continued

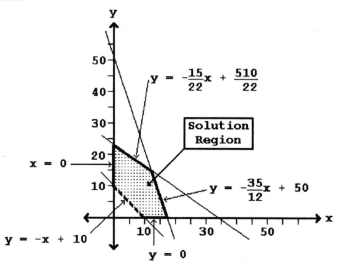

b)   The region of solutions is bounded.

c)   There are five corner points. Starting at the origin $(0, 0)$
     move counter-clockwise around the region of solutions.

P_1:   Intersection of $x = 0$ and $y = -x + 10$.

Substituting $x = 0$ into $y = -x + 10$ gives

$$y = -(0) + 10 = 10$$

$P_1 = (0, 10)$

P_2:   Intersection of $x = 0$ and $y = -\dfrac{15}{22}x + \dfrac{510}{22}$.

Substituting $x = 0$ into $y = -\dfrac{15}{22}(0) + \dfrac{510}{22} = \dfrac{510}{22} = 23\dfrac{2}{11}$

$P_2 = \left(0,\ 23\dfrac{2}{11}\right)$

17.c) Continued

P_3: Intersection of $y = -\frac{15}{22}x + \frac{510}{22}$ and $y = -\frac{35}{12}x + 50$

Multiply both equations by $22(12) = 264$

$264y = -180x + 6,120 \;\rightarrow$

$264y = -770x + 13,200 \;\rightarrow$

$$264y = -180x + 6,120$$
$$\underline{-264y = 770x - 13,200}$$
$$0 = 590x - 7,080$$

$$-590x = -7,080$$

$$x = \frac{7,080}{590} = 12$$

$$y = -\frac{35}{12}x + 50 = -\frac{35}{12}(12) + 50 = -35 + 50 = 15$$

$P_3 = (12,\ 15)$

P_4: Intersection of $y = 0$ and $y = -\frac{35}{12}x + 50.$

Substituting $y = 0$ into $y = -\frac{35}{12}(x) + 50$ gives

$$0 = -\frac{35}{12}x + 50 \;\rightarrow\; \frac{35}{12}x = 50 \;\rightarrow\; 35x = 50(12)$$

$$x = \frac{600}{35} = 17\tfrac{1}{7}$$

$P_4 = \left(17\tfrac{1}{7},\ 0\right)$

P_5: Intersection of $y = 0$ and $y = -x + 10.$

Substituting $y = 0$ into $y = -x + 10$ gives

$$0 = -x + 10 \;\rightarrow\; x = 10$$

$P_5 = (10,\ 0)$

Exercise 8.0

21.   a)

Original Inequality ········ Slope-Intercept Form	Associated Equation	Graph of the Inequality
$x - 2y + 16 \geq 0$ ········ $-2y \geq -x - 16 \quad \rightarrow$ $y \leq \frac{1}{2}x + 8$	$y = \frac{1}{2}x + 8$	all points on or below the line with slope (1/2) and y-intercept 8
$3x + y \leq 30$ ········ $y \leq -3x + 30$	$y = -3x + 30$	all points on or below the line with slope -3 and y-intercept 30
$x + y \leq 14$ ········ $y \leq -x + 14$	$y = -x + 14$	all points on or below the line with slope -1 and y-intercept 14
$x \geq 0$ ········ (not applicable)	$x = 0$	all points on or to the right of the y-axis
$y \geq 0$ ········ $y \geq 0$	$y = 0$	all points on or above the x-axis

21.a)   Continued

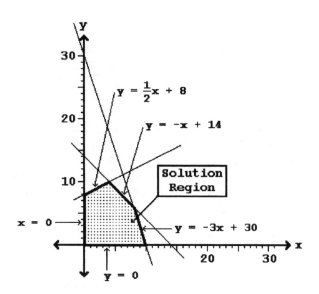

b)   The region of solutions is bounded.

c)   There are five corner points. Starting at the origin (0, 0)
     move counter-clockwise around the region of solutions.

P_1: Intersection of x = 0 and y = 0. P_1 = (0, 0)

P_2: The y-intercept of $y = \frac{1}{2}x + 8$.

   P_2 = (0, 8)

P_3: Intersection of $y = \frac{1}{2}x + 8$ and y = -x + 14.

   Multiply both equations by the LCD of 2.

   $\begin{aligned} 2y &= 1x + 16 \\ 2y &= -2x + 28 \end{aligned}$ $\rightarrow$ $\begin{aligned} 2y &= 1x + 16 \\ \underline{-2y} &= \underline{2x - 28} \\ 0 &= 3x - 12 \end{aligned}$

   $-3x = -12$

   $x = \dfrac{-12}{-3} = 4$

   y = -x + 14 = -(4) + 14 = 10

   P_3 = (4, 10)

21. Continued

P_4: Intersection of $y = -x + 14$ and $y = -3x + 30$.

$$
\begin{array}{ll}
y = -1x + 14 \rightarrow & \quad y = -1x + 14 \\
y = -3x + 30 \rightarrow & \underline{\quad -y = 3x - 30} \\
& \quad0 = 2x - 16
\end{array}
$$

$$-2x = -16$$

$$x = 8$$

$y = -x + 14 = -(8) + 14 = 6$

$P_4 = (8, 6)$

P_5: Intersection of $y = 0$ and $y = -3x + 30$.

Substituting $y = 0$ into $y = -3x + 30$ gives

$$0 = -3x + 30 \quad \rightarrow 3x = 30 \quad \rightarrow \quad x = 10$$

$P_5 = (10, 0)$

25 to 45. Omitted. See exercises 1 to 21. Adjust the viewing window on your calculator if needed for a particular exercise.

Exercise 8.1

1. Independent variables:
   x = number of shrubs
   y = number of trees

   Constraint:

$$
\begin{array}{c}
\text{gallons of water used} \le 100 \\
\text{(water for shrubs) + (water for trees)} \le 100 \\
\text{(1 gallon per shrub)(number of shrubs) +} \\
\text{(3 gallons per tree)(number of trees)} \le 100
\end{array}
$$

$$1x + 3y \le 100$$

5.   Independent variables:
          x = number of refrigerators
          y = number of dishwashers

      Constraint:

                                    warehouse storage space $\leq$ 1650
             (space for refrigerators) + (space for dishwashers) $\leq$ 1650
      (63 cubic feet per refrigerator)·(number of refrigerators)
        + (41 cubic feet per dishwasher)·(number of dishwashers) $\leq$ 1650

                                              $63x + 41y \leq 1650$

9.   **Step 1:**   *List the independent variables*

      Independent variables:
          x = pounds of Morning Blend to be prepared each day
          y = pounds of South American Blend prepared each day

      Step 2:   *List the constraints as linear inequalities*

      Constraints:

                                            Mexican beans used $\leq$ 100
      (Mexican in Morning)   + (Mexican in South American) $\leq$ 100
      (one-third of Morning) + (two-thirds of South American) $\leq$ 100

                                              $\frac{1}{3}x + \frac{2}{3}y \leq 100$

                                            Columbian beans used $\leq$ 80
      (Columbian in Morning)   + (Columbian in South American) $\leq$ 80
      (two-thirds of Morning) + (one-third of South American) $\leq$ 80

                                              $\frac{2}{3}x + \frac{1}{3}y \leq 80$

      Pounds of coffee cannot be negative, so $x \geq 0$ and $y \geq 0$.

      Step 3:   *Find the objective function (maximize profit)*

          z = (Morning profit) + (South American profit)
            = ($3 per pound)(pounds of Morning)
                + ($2.50 per pound)(pounds of South American)

          z = 3x + 2.5y

9. Continued

Step 4:   *Graph the region of possible solutions*

Original Inequality	Associated Equation	Graph of the Inequality
Slope-Intercept Form		
$\frac{1}{3}x + \frac{2}{3}y \leq 100$ $\frac{2}{3}y \leq -\frac{1}{3}x + 100 \quad \rightarrow$ $y \leq -\frac{1}{2}x + 150$	$y = -\frac{1}{2}x + 150$	all points on or below the line with slope $-(1/2)$ and y-intercept 150 and x-intercept 300
$\frac{2}{3}x + \frac{1}{3}y \leq 80$ $\frac{1}{3}y \leq -\frac{2}{3}x + 80 \quad \rightarrow$ $y \leq -2x + 240$	$y = -2x + 240$	all points on or below the line with slope -2 and y-intercept 240 and x-intercept 120
$x \geq 0$ (not applicable)	$x = 0$	all points on or to the right of the y-axis
$y \geq 0$ $y \geq 0$	$y = 0$	all points on or above the x-axis

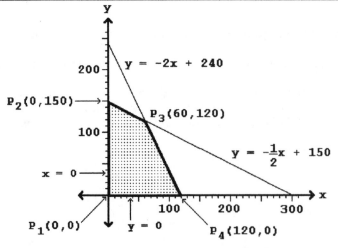

9.   Continued

Step 5: *Find all corner points and their z-values*

Use the elimination method to find P_3.

$$\frac{1}{3}x + \frac{2}{3}y = 100 \quad \rightarrow \quad x + 2y = 300 \quad \rightarrow \quad -2x - 4y = -600$$

$$\frac{2}{3}x + \frac{1}{3}y = 80 \quad \rightarrow \quad 2x + y = 240 \quad \rightarrow \quad \underline{2x + y = 240}$$

$$-3y = -360$$
$$y = 120$$

$$2x + (120) = 240$$
$$2x = 120$$
$$x = 60$$

The corner points are: $P_1(0,0)$, $P_2(0,150)$, $P_3(60,120)$, $P_4(120,0)$

Point	Value of z = 3x + 2.5y
$P_1(0,0)$	z = 3(0) + 2.5(0) = 0
$P_2(0,150)$	z = 3(0) + 2.5(150) = 375
$P_3(60,120)$	z = 3(60) + 2.5(120) = 480
$P_4(120,0)$	z = 3(120) + 2.5(0) = 360

Step 6: *Find the maximum profit*

Pete's Coffees should produce 60 pounds of Morning Blend and 120 pounds of South American Blend each day to maximize their profit at $480.

Exercise 8.1

13.

	First Class Passengers	Economy Passengers	Cost	Maximum Flights
Orville 606	20	80	12,000	52
Wilbur W-1112	80	120	18,000	30
Minimum Needed	1600	4800		

Step 1: *List the independent variables*

Independent variables:
    x = flights using Orville 606
    y = flights using Wilbur W-1112

Step 2: *List the constraints as linear inequalities*

Constraints:
$$\text{First class passengers} \geq 1600$$
(First class on Orville) + (First class on Wilbur) $\geq$ 1600
        (20 on Orville)•(flights of Orville) +
                (80 on Wilbur)•(flights of Wilbur) $\geq$ 1600

$$20x + 80y \geq 1600$$

$$\text{Economy passengers} \geq 4800$$
(Economy on Orville) + (Economy on Wilbur) $\geq$ 4800
    (80 on Orville)•(flights of Orville) +
            (120 on Wilbur)•(flights of Wilbur) $\geq$ 4800

$$80x + 120y \geq 4800$$

Number of flights on Orville $\leq$ 52
$$x \leq 52$$

Number of flights on Wilbur $\leq$ 30
$$y \leq 30$$

Number of flights cannot be negative, so $x \geq 0$ and $y \geq 0$.

Step 3: *Find the objective function (minimize cost)*

    z = (Orville cost) + (Wilbur cost)
      = ($12,000 per flight)•(flights of Orville)
          + ($18,000 per flight)•(flights of Wilbur)

    z = 12,000x + 18,000y

13. Continued

Step 4: *Graph the region of possible solutions*

Original Inequality Slope-Intercept Form	Associated Equation	Graph of the Inequality
$20x + 80y \geq 1600$ $80y \geq -20x + 1600 \rightarrow$ $y \geq -\frac{1}{4}x + 20$	$y = -\frac{1}{4}x + 20$	all points on or below the line with slope $-(1/4)$ and y-intercept 20 and x-intercept 80
$80x + 120y \geq 4800$ $120y \geq -80x + 4800 \rightarrow$ $y \geq -\frac{2}{3}x + 40$	$y = -\frac{2}{3}x + 40$	all points on or below the line with slope $-(2/3)$ and y-intercept 40 and x-intercept 60
$x \leq 52$ (not applicable)	$x = 52$	all points on or to the left of the vertical line through x = 52
$y \leq 30$ $y \leq 30$	$y = 30$	all points on or above the horizontal line through y = 30
$x \geq 0$ (not applicable)	$x = 0$	all points on or to the right of the y-axis
$y \geq 0$ $y \geq 0$	$y = 0$	all points on or above the x-axis

13. Continued

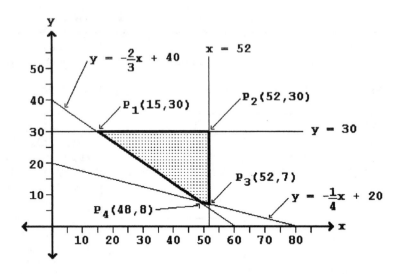

Step 5: *Find all corner points and their z-values*

Use the elimination method to find P_4.

$$20x + 80y = 1600 \quad \rightarrow \quad -80x - 320y = -6400$$
$$80x + 120y = 4800 \quad \rightarrow \quad \underline{80x + 120y = 4800}$$
$$-200y = -1600$$
$$y = 8$$

$$20x + 80(8) = 1600$$
$$20x = 1600 - 640$$
$$20x = 960$$
$$x = 48$$

The corner points are: $P_1(15,30)$, $P_2(52,30)$, $P_3(52,7)$, $P_4(48,8)$

Point	Value of $z = 12,000x + 18,000y$
$P_1(15,30)$	$z = 12,000(15) + 18,000(30) = 720,000$
$P_2(52,30)$	$z = 12,000(52) + 18,000(30) = 1,164,000$
$P_3(52,7)$	$z = 12,000(52) + 18,000(7) = 750,000$
$P_4(48,8)$	$z = 12,000(48) + 18,000(8) = 720,000$

13. Continued

    Step 6: *Find the minimum cost*

    To minimize its costs Global Air Lines should schedule between 15 and 48 flights on Orville 606 and between 8 and 30 flights on Wilbur W-1112 such that they satisfy the equation L_2: $y = -(2/3)x + 40$. The operating costs for using that type of scheduling would be \$720,000. Some specific solutions might be:

    15 Orvilles and 30 Wilburs
    18 Orvilles and 28 Wilburs
    21 Orvilles and 26 Wilburs

17. Mexican beans used = $\frac{1}{3}$(Morning Blend) + $\frac{2}{3}$(South American)

    $= \frac{1}{3}$(60 pounds) + $\frac{2}{3}$(120 pounds)

    $= 20 + 80 = 100$ pounds

    Mexican unused = Mexican available − Mexican used
    $= 100 - 100 = 0$ pounds

    Columbian beans used = $\frac{2}{3}$(Morning Blend) + $\frac{1}{3}$(South American)

    $= \frac{2}{3}$(60 pounds) + $\frac{1}{3}$(120 pounds)

    $= 40 + 40 = 80$ pounds

    Columbian unused = Columbian available − Columbian used
    $= 80 - 80 = 0$ pounds

21. Omitted.

25. Omitted.

29. Omitted.

Exercise 8.1

33.

	Quarter-ton	Half-ton	Three-quarter-ton	Weekly Cost
Detroit	30	60	90	540,000
Los Angeles	60	45	30	360,000
Minimum Needed	300	450	450	

Step 1: *List the independent variables*

Independent variables:
  x = number of weeks needed at Detroit plant
  y = number of weeks needed at Los Angeles plant

Step 2: *List the constraints as linear inequalities*

Constraints:
                    Quarter-ton trucks needed $\geq$ 300
(Quarter-tons at Detroit) + (Quarter-tons at Los Angeles) $\geq$ 300
(30 per week at Detroit)•(weeks at Detroit) +
     (60 per week at Los Angeles)•(weeks at Los Angeles) $\geq$ 300

$$30x + 60y \geq 300$$

                    Half-ton trucks needed $\geq$ 450
(Half-tons at Detroit) + (Half-tons at Los Angeles) $\geq$ 450
(60 per week at Detroit)•(weeks at Detroit) +
  (45 per week at Los Angeles)•(weeks at Los Angeles) $\geq$ 450

$$60x + 45y \geq 450$$

                    Three-quarter-ton trucks needed $\geq$ 450
(Three-Qtr-tons at Detroit) + (Three-Qtr-tons at L.A.) $\geq$ 450
(90 per week at Detroit)•(weeks at Detroit) +
   (30 per week at Los Angeles)•(weeks at Los Angeles) $\geq$ 450

$$90x + 30y \geq 450$$

Number of weeks cannot be negative, so $x \geq 0$ and $y \geq 0$.

Step 3: *Find the objective function (minimize cost)*

    z = (Detroit cost) + (Los Angeles cost)
      = ($540,000 per week)•(weeks at Detroit)
        + ($360,000 per week)•(weeks at Los Angles)

    $z = 540,000x + 360,000y$

33. Continued

Step 4: *Graph the region of possible solutions*

Original Inequality	Associated Equation	Graph of the Inequality
Slope-Intercept Form		
$30x + 60y \geq 300$ $60y \geq -30x + 300 \rightarrow$ $y \geq -\frac{1}{2}x + 5$	$y = -\frac{1}{2}x + 5$	all points on or above the line with slope $-(1/2)$ and y-intercept 5 and x-intercept 10
$60x + 45y \geq 450$ $45y \geq -60x + 450 \rightarrow$ $y \geq -\frac{4}{3}x + 10$	$y = -\frac{4}{3}x + 10$	all points on or above the line with slope $-(4/3)$ and y-intercept 10 and x-intercept 7.5
$90x + 30y \geq 450$ $30y \geq -90x + 450 \rightarrow$ $y \geq -3x + 15$	$y = -3x + 15$	all points on or above the line with slope -3 and y-intercept 15 and x-intercept 5
$x \geq 0$ (not applicable)	$x = 0$	all points on or to the right of the y-axis
$y \geq 0$ $y \geq 0$	$y = 0$	all points on or above the x-axis

Exercise 8.1

33. Continued

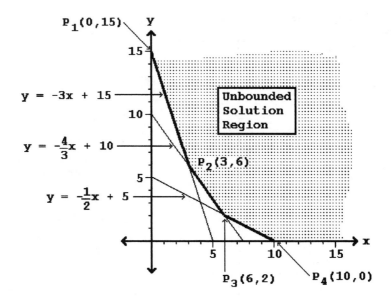

P₁(0,15)
y = -3x + 15
y = -⁴⁄₃x + 10
Unbounded Solution Region
y = -½x + 5
P₂(3,6)
P₃(6,2)
P₄(10,0)

Step 5: *Find all corner points and their z-values*

Use the elimination method to find P_2.

$$y = -\frac{4}{3}x + 10 \quad \text{(multiply by 3)} \quad \rightarrow \quad 3y = -4x + 30$$

$$y = -3x + 15 \quad \text{(multiply by -3)} \quad \rightarrow \quad \underline{-3y = 9x - 45}$$
$$0 = 5x - 15$$
$$-5x = -15$$
$$x = 3$$

$$y = -3x + 15 = -3(3) + 15 = -9 + 15 = 6$$
$$P_2 = (3,6)$$

Use the elimination method to find P_3.

$$y = -\frac{1}{2}x + 5 \quad \text{(multiply by 6)} \quad \rightarrow \quad 6y = -3x + 30$$

$$y = -\frac{4}{3}x + 10 \quad \text{(multiply by -6)} \quad \rightarrow \quad \underline{-6y = 8x - 60}$$
$$0 = 5x - 30$$
$$-5x = -30$$
$$x = 6$$

$$y = -\frac{1}{2}x + 5 = -\frac{1}{2}(6) + 5 = -3 + 5 = 2$$
$$P_3 = (6,2)$$

33. Continued

The corner points are: $P_1(0,15)$, $P_2(3,6)$, $P_3(6,2)$, $P_4(10,0)$

Point	Value of z = 540,000x + 360,000y
$P_1(0,15)$	z = 540,000(0) + 360,000(15) = 5,400,000
$P_2(3,6)$	z = 540,000(3) + 360,000(6) = 3,780,000
$P_3(6,2)$	z = 540,000(6) + 360,000(2) = 3,960,000
$P_4(10,0)$	z = 540,000(10) + 360,000(0) = 5,400,000

Step 6: *Determine if a maximum and/or minimum exists*

Let z equal various values and solve for y.

z = 3,780,000

$$540,000x + 360,000y = 3,780,000$$
$$360,000y = -540,000x + 3,780,000$$

$$y = -\frac{3}{2}x + \frac{21}{2}$$

slope = -3/2, y-intercept = 10.5, x-intercept = 7

z = 5,400,000

$$540,000x + 360,000y = 3,780,000$$
$$360,000y = -540,000x + 3,780,000$$

$$y = -\frac{3}{2}x + 15$$

slope = -3/2, y-intercept = 15, x-intercept = 10

Graph these lines (which are parallel) to show the relative values of z. The closer the line is to the upper right the larger the value of z, and the closer the line is to the origin the smaller the value of z.

33.   Continued

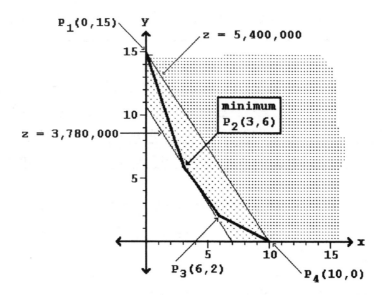

P$_1$(0,15)   y

z = 5,400,000

15

minimum
P$_2$(3,6)

10

z = 3,780,000

5

x

5   10   15

P$_3$(6,2)

P$_4$(10,0)

$P_2(3,6)$ is the point at which the minimum occurs. Note the line
labeled z = 3,780,000 matches the solution region at only this
point.   The minimum value is 3,780,000.

Exercise 8.2

1.   The inequality is: $3x_1 + 2x_2 \leq 5$
The equation is: $3x_1 + 2x_2 + s_1 = 5$

5.   a)   The inequality is: $6.99x_1 + 3.15x_2 + 1.98x_3 \leq \$42.15$
b)   The equation is: $6.99x_1 + 3.15x_2 + 1.98x_3 + s_1 = \$42.15$
c) x_1 = pounds of meat
x_2 = pounds of cheese
x_3 = loaves of bread
s_1 = unused money after the purchases have been made

9.   a)   The inequality is: $24x_1 + 36x_2 + 56x_3 + 72x_4 \leq 50,000$
b)   The equation is: $24x_1 + 36x_2 + 56x_3 + 72x_4 + s_1 = 50,000$
c) x_1 = number of twin beds
x_2 = number of double beds
x_3 = number of queen-size beds
x_4 = number of king-size beds
s_1 = cubic feet of unused space in the mattress
warehouse after the 4 types of beds are stored

13. Non-identity matrix columns (value is 0): $x_1 = 0$, $x_2 = 0$

    Identity matrix columns (turn at 1 and read last column for answer): $s_1 = 7.8$, $s_2 = 9.3$, $s_3 = 0.5$, $z = 9.6$

    So $(x_1, x_2, s_1, s_2, s_3) = (0, 0, 7.8, 9.3, 0.5)$ with $z = 9.6$

17. Constraint equations:
    $$C_1: \quad 3x_1 + 4x_2 + s_1 = 40$$
    $$C_2: \quad 4x_1 + 7x_2 + s_2 = 50$$
    $$(x_1, x_2, s_1, s_2 \geq 0)$$

    Objective function equation:
    $z = 2x_1 + 4x_2$ becomes
    $-2x_1 - 4x_2 + 0s_1 + 0s_2 + 1z = 0$

    First simplex matrix:

x_1	x_2	s_1	s_2	z	
3	4	1	0	0	40
4	7	0	1	0	50
-2	-4	0	0	1	0

    Possible solution:
    $(x_1, x_2, s_1, s_2) = (0, 0, 40, 50)$ with $z = 0$

21. Constraint equations:
    $$C_1: \quad 5x_1 + 3x_2 + 9x_3 + s_1 = 10$$
    $$C_2: \quad 12x_1 + 34x_2 + 100x_3 + s_2 = 10$$
    $$C_3: \quad 52x_1 + 7x_2 + 12x_3 + s_3 = 10$$
    $$(x_1, x_2, x_3, s_1, s_2, s_3 \geq 0)$$

    Objective function equation:
    $z = 4x_1 + 7x_2 + 9x_3$ becomes
    $-4x_1 - 7x_2 - 9x_3 + 0s_1 + 0s_2 + 0s_3 + 1z = 0$

    First simplex matrix:

x_1	x_2	x_3	s_1	s_2	s_3	z	
5	3	9	1	0	0	0	10
12	34	100	0	1	0	0	10
52	7	12	0	0	1	0	10
-4	-7	-9	0	0	0	1	0

    Possible solution:
    $(x_1, x_2, x_3, s_1, s_2, s_3) = (0, 0, 0, 10, 10, 10)$ with $z = 0$

Exercise 8.2

25. **Step 1:** *Model the problem*

    List the independent variables

    Independent variables:
    x_1 = pounds of Yusip Blend to be prepared each day
    x_2 = pounds of Exotic Blend prepared each day

    List the constraints as linear inequalities

    Constraints:
    Costa Rican beans used $\leq$ 200
    (Costa Rican in Yusip) + (Costa Rican in Exotic) $\leq$ 200
    (one-half of Yusip) + (one-quarter of Exotic) $\leq$ 200

    C_1: $\frac{1}{2}x_1 + \frac{1}{4}x_2 \leq 200$

    Ethiopian beans used $\leq$ 330
    (Ethiopian in Yusip) + (Ethiopian in Exotic) $\leq$ 330
    (one-half of Yusip) + (three-quarters of Exotic) $\leq$ 330

    C_2: $\frac{1}{2}x_1 + \frac{3}{4}x_2 \leq 330$

    Pounds of coffee cannot be negative, so $x_1 \geq 0$ and $x_2 \geq 0$.

    Find the objective function (maximize profit)

    z = (Yusip profit) + (Exotic profit)
    = ($3.50 per pound)(pounds of Yusip)
    + ($4 per pound)(pounds of Exotic)

    $z = 3.5x_1 + 4x_2$

    Step 2: *Convert each constraint to an equation*

    C_1: $\frac{1}{2}x_1 + \frac{1}{4}x_2 + s_1 = 200$ where s_1 = unused Costa Rican beans

    C_2: $\frac{1}{2}x_1 + \frac{3}{4}x_2 + s_2 = 330$ where s_2 = unused Ethiopian beans

    $(x_1, x_2, s_1, s_2 \geq 0)$

    Step 3: *Rewrite the objective function*

    $z = 3.5x_1 + 4x_2$ becomes
    $-3.5x_1 - 4x_2 + 0s_1 + 0s_2 + 1z = 0$

25. Continued

Step 4: *First simplex matrix*

$$
\begin{array}{ccccc}
x_1 & x_2 & s_1 & s_2 & z \\
\end{array}
$$

$$
\left[
\begin{array}{ccccc|c}
1/2 & 1/4 & 1 & 0 & 0 & 200 \\
1/2 & 3/4 & 0 & 1 & 0 & 330 \\
-3.5 & -4 & 0 & 0 & 1 & 0
\end{array}
\right]
$$

Step 5: *Find the possible solution*

$$(x_1, x_2, s_1, s_2) = (0, 0, 200, 330) \text{ with } z = 0$$

Exercise 8.3

1.
$$
\begin{array}{ccccc}
x_1 & x_2 & s_1 & s_2 & z \\
\end{array}
$$

$$
\left[
\begin{array}{ccccc|c}
5 & 0 & 3 & 1 & 0 & 12 \\
3 & 0 & \mathbf{19} & 0 & 10 & 22 \\
-21 & 1 & -48 & 0 & 0 & 19
\end{array}
\right]
\begin{array}{l}
\leftarrow 12/3 = 4 \\
\leftarrow 22/19 \approx 1.2 \quad \leftarrow \text{ pivot row} \\
\leftarrow \text{ not a constraint row}
\end{array}
$$

$$\uparrow$$
pivot column

The most negative entry in the last row is -48 in column 3. Divide the last entry in each constraint row by the corresponding entry in column 3. The pivot row is row 2 which yielded the smallest positive quotient. The pivot entry is 19 in row 2, column 3.

5. a)
$$
\begin{array}{ccccc}
x_1 & x_2 & s_1 & s_2 & z \\
\end{array}
$$

$$
\left[
\begin{array}{ccccc|c}
1 & 2 & 1 & 0 & 0 & 3 \\
\mathbf{4} & 1 & 0 & 1 & 0 & 2 \\
-6 & -4 & 0 & 0 & 1 & 0
\end{array}
\right]
\begin{array}{l}
\leftarrow 3/1 = 3 \\
\leftarrow 2/4 = 0.5 \quad \leftarrow \text{ pivot row} \\
\leftarrow \text{ not a constraint row}
\end{array}
$$

$$\uparrow$$
pivot column

The most negative entry in the last row is -6 in column 1. Divide the last entry in each constraint row by the corresponding entry in column 1. The pivot row is row 2 which yielded the smallest positive quotient. The pivot entry is 4 in row 2, column 1.

Exercise 8.3

5.   Continued

b)

$$R2 \div 4 : R2 \begin{bmatrix} 1 & 2 & 1 & 0 & 0 & 3 \\ 1 & 1/4 & 0 & 1/4 & 0 & 1/2 \\ -6 & -4 & 0 & 0 & 1 & 0 \end{bmatrix}$$

$$\begin{matrix} -R2 + R1 : R1 \\ \\ 6R2 + R3 : R3 \end{matrix} \begin{bmatrix} 0 & 7/4 & 1 & -1/4 & 0 & 5/2 \\ 1 & 1/4 & 0 & 1/4 & 0 & 1/2 \\ 0 & -5/2 & 0 & 3/2 & 1 & 3 \end{bmatrix}$$

c) $(x_1, x_2, s_1, s_2) = \left(\dfrac{1}{2}, \ 0, \ 2\dfrac{1}{2}, \ 0 \right)$ with $z = 3$

9.   **Step 1:**   *Model the problem*

Independent variables:
        x_1 = loaves of bread produced each day
        x_2 = number of cakes produced each day

Constraints:
        C_1:   Time
            5 friends(8hr)(60min/hr) = 2400 minutes per day
                                Working time per day $\leq$ 2400
                    (bread making time) + (cake making time) $\leq$ 2400
            (time per loaf)(loaves) + (time per cake)(cakes) $\leq$ 2400
                                            $50x_1 + 30x_2 \leq 2400$

        C_2:   Cost
                                            Daily costs $\leq$ 190
                            (bread cost) + (cake cost) $\leq$ 190
            (cost per loaf)(loaves) + (cost per cake)(cakes) $\leq$ 190
                                            $0.90x_1 + 1.50x_2 \leq 190$

Objective Function:
        z = (bread profit) + (cake profit)
          = ($1.20 per loaf)(loaves of bread)
            + ($4.00 per cake)(number of cakes)

        $z = 1.2x_1 + 4x_2$

Step 2:   *Find the constraint equations*

        C_1: $50x_1 + 30x_2 + s_1 = 2,400$ where s_1 = unused time
        C_2: $0.9x_1 + 1.5x_2 + s_2 = 190$ where s_2 = unused money

9. Continued

Step 3: *Rewrite the objective function equation*

$z = 1.2x_1 + 4x_2$ becomes
$-1.2x_1 - 4x_2 + 0s_1 + 0s_2 + 1z = 0$

Step 4: *Find the first simplex matrix*

$$
\begin{array}{ccccc}
x_1 & x_2 & s_1 & s_2 & z
\end{array}
$$

$$
\begin{bmatrix}
50 & 30 & 1 & 0 & 0 & 2400 \\
0.9 & 1.5 & 0 & 1 & 0 & 190 \\
-1.2 & -4 & 0 & 0 & 1 & 0
\end{bmatrix}
$$

Step 5: *Find the possible solution*

$(x_1, x_2, s_1, s_2) = (0, 0, 2400, 190)$ with $z = 0$

Step 6: *Pivot to find a better possible solution*

$$
\begin{array}{ccccc}
x_1 & x_2 & s_1 & s_2 & z
\end{array}
$$

$$
\begin{bmatrix}
50 & \mathbf{30} & 1 & 0 & 0 & 2400 \\
0.9 & 1.5 & 0 & 1 & 0 & 190 \\
-1.2 & -4 & 0 & 0 & 1 & 0
\end{bmatrix}
\begin{array}{l}
\leftarrow 2400/30 = 80 \leftarrow \text{pivot row} \\
\leftarrow 190/1.5 \approx 126.7 \\
\leftarrow \text{not a constraint row}
\end{array}
$$

$\uparrow$
pivot column

$$
R1 \div 30{:}R1
\begin{bmatrix}
1.6667 & 1 & 0.0333 & 0 & 0 & 80 \\
0.9 & 1.5 & 0 & 1 & 0 & 190 \\
-1.2 & -4 & 0 & 0 & 1 & 0
\end{bmatrix}
$$

$$
\begin{array}{ccccc}
x_1 & x_2 & s_1 & s_2 & z
\end{array}
$$

$$
\begin{array}{l}
\\
-1.5R1 + R2{:}R2 \\
4.00R1 + R3{:}R3
\end{array}
\begin{bmatrix}
1.6667 & 1 & 0.0333 & 0 & 0 & 80 \\
-1.6 & 0 & -0.050 & 1 & 0 & 70 \\
5.4667 & 0 & 0.1333 & 0 & 1 & 320
\end{bmatrix}
$$

Step 7: *Determine the possible solution*

$(x_1, x_2, s_1, s_2) = (0, 80, 0, 70)$ with $z = 320$

Check:
C_1: $50x_1 + 30x_2 + s_1 = 50(0) + 30(80) + 0 = 2,400$
C_2: $0.9x_1 + 1.5x_2 + s_2 = 0.9(0) + 1.5(80) + 70 = 190$
$z = 1.2x_1 + 4x_2 = 1.2(0) + 4(80) = 320$

9.   Continued

Step 8: *Determine if this maximizes the objective function*

The last row contains no negative entries, so we do not pivot again.

Step 9: *Interpret the final solution*

The five friends should make 80 cakes each day and no bread. This results in no unused hours and $70 in unused money. Their maximum profit will be $320.

13. **Step 1:** *Model the problem*

Independent variables:
x_1 = liters of House White wine produced each day
x_2 = liters of Premium White wine produced each day
x_3 = liters of Sauvignon Blanc wine produced each day

Constraints:
2 pounds of grapes = 1 liter of wine

C_1:   French Colombard grapes
30,000 pounds of grapes = 15,000 liters of wine
French Colombard (FC) available $\leq$ 15,000
(FC in House) + (FC in Premium) + (FC in Sauvignon) $\leq$ 15,000
(75% of House) + (25% of Premium) + (0% of Sauvi) $\leq$ 15,000
$0.75x_1 + 0.25x_2 + 0x_3 \leq 15,000$

C_2:   Sauvignon Blanc grapes
20,000 pounds of grapes = 10,000 liters of wine
Sauvignon Blanc (SB) available $\leq$ 10,000
(SB in House) + (SB in Premium) + (SB in Sauvignon) $\leq$ 10,000
(25% of House) + (75% of Premium) + (100% of Sauvi) $\leq$ 10,000
$0.25x_1 + 0.75x_2 + 1x_3 \leq 10,000$

Objective Function:
z = (House profit) + (Premium profit) + (Sauvignon profit)
= ($1 per liter)(liters of House)
+ ($1.50 per liter)(liters of Premium)
+ ($2 per liter)(liters of Sauvignon Blanc)

$z = 1x_1 + 1.5x_2 + 2x_3$

13. Continued

    Step 2: *Find the constraint equations*

    C_1: $0.75x_1 + 0.25x_2 + 0x_3 + s_1 = 15,000$ where $s_1 =$ unused FC grapes

    C_2: $0.25x_1 + 0.75x_2 + 1x_3 + s_2 = 10,000$ where $s_2 =$ unused SB grapes

    Step 3: *Rewrite the objective function equation*

    $z = 1x_1 + 1.5x_2 + 2x_3$ becomes
    $-1x_1 - 1.5x_2 - 2x_3 + 0s_1 + 0s_2 + 1z = 0$

    Step 4: *Find the first simplex matrix*

x_1	x_2	x_3	s_1	s_2	z	
0.75	0.25	0	1	0	0	15,000
0.25	0.75	1	0	1	0	10,000
−1.0	−1.5	−2	0	0	1	0

    Step 5: *Find the possible solution*

    $(x_1, x_2, x_3, s_1, s_2) = (0,0,0, 15,000, 10,000)$ with $z = 0$

    Step 6: *Pivot to find a better possible solution*

x_1	x_2	x_3	s_1	s_2	z		
0.75	0.25	0	1	0	0	15,000	Undefined
0.25	0.75	**1**	0	1	0	10,000	10,000 ← pivot row
−1.0	−1.5	−2	0	0	1	0	

    $\uparrow$
    pivot column

    No division necessary

	0.75	0.25	0	1	0	0	15,000
	0.25	0.75	1	0	1	0	10,000
2R2 + R3:R3	−0.5	0	0	0	2	1	20,000

13. Continued

Last row is not all positive, so pivot again.

$$\begin{bmatrix} \mathbf{0.75} & 0.25 & 0 & 1 & 0 & 0 & 15,000 \\ 0.25 & 0.75 & 1 & 0 & 1 & 0 & 10,000 \\ -0.5 & 0 & 0 & 0 & 2 & 1 & 20,000 \end{bmatrix} \begin{matrix} 20,000 \leftarrow \text{pivot row} \\ 40,000 \\ \end{matrix}$$

$\uparrow$
pivot column

$$\text{R1} \div 0.75 \text{:R1} \begin{bmatrix} 1 & 0.3333 & 0 & 1.3333 & 0 & 0 & 20,000 \\ 0.25 & 0.75 & 1 & 0 & 1 & 0 & 10,000 \\ -0.5 & 0 & 0 & 0 & 2 & 1 & 20,000 \end{bmatrix}$$

$$\begin{matrix} & x_1 & x_2 & x_3 & s_1 & s_2 & z \end{matrix}$$

$$\begin{matrix} \phantom{-0.25\text{R1 + R2:R2}} \\ -0.25\text{R1} + \text{R2:R2} \\ 0.5\text{R1} + \text{R3:R3} \end{matrix} \begin{bmatrix} 1 & 0.3333 & 0 & 1.3333 & 0 & 0 & 20,000 \\ 0 & 0.6667 & 1 & -0.333 & 1 & 0 & 5,000 \\ 0 & 0.1667 & 0 & 0.6667 & 2 & 1 & 30,000 \end{bmatrix}$$

Last row is all positive, so we do not need to pivot again.

Step 7: *Determine the possible solution*

$(x_1, x_2, x_3, s_1, s_2) = (20,000, 0, 5000, 0, 0)$ with $z = 30,000$

Check:

C_1: $0.75x_1 + 0.25x_2 + 0x_3 + s_1$
  $= 0.75(20,000) + 0.25(0) + 0(5000) + 0 = 15,000$

C_2: $0.25x_1 + 0.75x_2 + 1x_3 + s_2$
  $= 0.25(20,000) + 0.75(0) + 1(5000) + 0$
  $= 5000 + 5000 = 10,000$

$z = 1x_1 + 1.5x_2 + 2x_3$
  $= 1(20,000) + 1.5(0) + 2(5000) = 20,000 + 10,000 = 30,000$

Step 8: *Determine if this maximizes the objective function*

The last row contains no negative entries, so we do not
pivot again.

13. Continued

    Step 9: *Interpret the final solution*

    J & M Winery should make 20,000 liters of House White wine, no Premium White and 5000 liters of Sauvignon Blanc wine to maximize its profit. The maximum profit would be $30,000. There would be no unused grapes of either the French Colombard variety or the Sauvignon Blanc variety.

17. Omitted.

21. Omitted

25. *Constraint equations*

$$C_1: \quad 6.32x_1 + 7.44x_2 + 8.32x_3 + 1.46x_4 + 9.35x_5 + s_1 = 63$$
$$C_2: \quad 8.36x_1 + 5.03x_2 + 1.00x_3 + 0x_4 + 5.25x_5 + s_2 = 32$$
$$C_3: \quad 1.14x_1 + 9.42x_2 + 9.39x_3 + 10.42x_4 + 9.32x_5 + s_3 = 14.7$$

    Rewrite the objective function equation

$$z = 37x_1 + 19x_2 + 53x_3 + 49x_4 \text{ becomes}$$
$$-37x_1 - 19x_2 - 53x_3 - 49x_4 - 0x_5 + 0s_1 + 0s_2 + 0s_3 + 1z = 0$$

    The first simplex matrix

x_1	x_2	x_3	x_4	x_5	s_1	s_2	s_3	z		
6.32	7.44	8.32	1.46	9.35	1	0	0	0	63	7.6
8.36	5.03	1.00	0	5.25	0	1	0	0	32	32
1.14	9.42	**9.39**	10.42	9.32	0	0	1	0	14.7	1.6 ← row
-37	-19	-53	-49	0	0	0	0	1	0	

$$\uparrow$$
$$\text{pivot column}$$

    The first possible solution

$$(x_1, x_2, x_3, x_4, x_5, s_1, s_2, s_3) = (0,0,0,0,0,63,32,14.7) \text{ with } z = 0$$

Exercise 8.3

25. Continued

Pivot to find the best possible solution

Matrix dimensions are 4 × 10

Pivot on 9.39 in row 3, column 3

Divide Row3 by 9.39 or Multiply by 1/9.39
Add −8.32•Row3 + Row1:Row1
Add −1.00•Row3 + Row2:Row2
Add +53.0•Row3 + Row4:Row4

Column 1 still has a negative in the last row, so pivot again.

    column 1   last column

$$\begin{bmatrix} 5.3099041533546 & \ldots & 49.9750 \\ \mathbf{8.2385942492013} & \ldots & 30.4345 \\ .12140575079872 & \ldots & 1.56549 \\ -30.56549520766 & \ldots & 82.9712 \end{bmatrix} \begin{matrix} 49.975/5.309 \approx 9.4 \\ 30.434/8.238 \approx 3.7 \leftarrow \text{pivot} \\ 1.5654/0.121 \approx 12.9 \\ \end{matrix}$$

Pivot on 8.23859... in Row 2, Column 1

Divide Row2 by 8.23859 or Multiply by 1/8.2385942492013
Add −5.3099...•Row2 + Row1:Row1
Add −0.1214...•Row2 + Row3:Row3
Add +30.565...•Row2 + Row4:Row4

There are no more negative numbers in the last row

Determine and check the best possible solution

$(x_1, x_2, x_3, x_4, s_1, s_2, s_3) = (3.694, 0, 1.117, 0, 0, 30.359, 0, 0)$
with $z = 195.884$

Check:
C_1: $6.32x_1 + 7.44x_2 + 8.32x_3 + 1.46x_4 + 9.35x_5 + s_1 = 63$
$= 6.32(3.694) + 0 + 8.32(1.117) + 0 + 0 + 30.359$
$= 23.34608 + 9.29344 + 30.359$
$= 62.99852 \approx 63$

C_2: $8.36x_1 + 5.03x_2 + 1.00x_3 + 0x_4 + 5.25x_5 + s_2 = 32$
$= 8.36(3.694) + 0 + 1.00(1.117) + 0 + 0 + 0$
$= 30.88184 + 1.117$
$= 31.99884 \approx 32$

25. Continued

C_3: $1.14x_1 + 9.42x_2 + 9.39x_3 + 10.42x_4 + 9.32x_5 + s_3 = 14.7$
$= 1.14(3.694) + 0 + 9.39(1.117) + 0 + 0 + 0$
$= 4.21116 + 10.48863$
$= 14.69979 \approx 14.7$

$z = 37x_1 + 19x_2 + 53x_3 + 49x_4$
$= 37(3.694) + 0 + 53(1.117) + 0$
$= 136.678 + 59.201$
$= 195.879 \approx 195.884$

29. **Step 1:** *Model the problem*

Independent variables:
x_1 = loaves of Nine-Grain Bread produced each week
x_2 = loaves of Sourdough Bread produced each week
x_3 = number of Chocolate Cakes produced each week
x_4 = number of Poppyseed Cakes produced each week
x_5 = dozens of Blueberry Muffins produced each week
x_6 = dozens of Apple-Cinnamon Muffins produced each week

Constraints:
C_1: Time
5 friends + 4 workers = 9 bakers
9 bakers(40 hr/wk)(60min/hr) = 21,600 minutes per week
Working time per week $\leq 21,600$
(time per each item)(number of items produced) $\leq 21,600$
$50x_1 + 50x_2 + 35x_3 + 30x_4 + 15x_5 + 15x_6 \leq 21,600$

C_2: Cost
Weekly costs $\leq 2,000$
(cost per each item)(number of items produced) $\leq 2,000$
$1.05x_1 + 0.95x_2 + 2.00x_3 + 1.55x_4 + 1.60x_5 + 1.30x_6 \leq 2,000$

Objective Function:
z = (profit per each item)(number of items produced)
$= 0.60x_1 + 0.70x_2 + 2.50x_3 + 2.00x_4 + 16.10x_5 + 14.40x_6$

Step 2: *Find the constraint equations*

C_1: $50x_1 + 50x_2 + 35x_3 + 30x_4 + 15x_5 + 15x_6 + s_1 = 21,600$
where s_1 = unused time
C_2: $1.05x_1 + 0.95x_2 + 2.00x_3 + 1.55x_4 + 1.60x_5 + 1.30x_6 + s_2$
$= 2,000$ where s_2 = unused money

Exercise 8.3

29. Continued

Step 3: *Rewrite the objective function equation*

$$z = 0.60x_1 + 0.70x_2 + 2.50x_3 + 2.00x_4 + 16.10x_5 + 14.40x_6$$
becomes

$$-0.60x_1 - 0.70x_2 - 2.50x_3 - 2.00x_4 - 16.10x_5 - 14.40x_6$$
$$+ 0s_1 + 0s_2 + 1z = 0$$

Step 4: *Find the first simplex matrix*

x_1	x_2	x_3	x_4	x_5	x_6	s_1	s_2	z		
50	50	35	30	15	15	1	0	0	21,600	1440
1.05	0.95	2.00	1.55	**1.60**	1.30	0	1	0	2,000	1250←row
-0.6	-0.7	-2.5	-2.0	-16.1	-14.4	0	0	1	0	

$$\uparrow$$
pivot column

Step 5: *Find the possible solution*

$$(x_1, x_2, x_3, x_4, x_5, x_6, s_2, s_3) = (0,\ 0,\ 0,\ 0,\ 0,\ 0,\ 21,600,\ 2,000)$$
$$\text{with } z = 0$$

Step 6: *Pivot to find the best possible solution*

Use the computer or calculator to find the best possible solution

Matrix dimensions are 3 × 10

Pivot on 1.60 in row 2, column 5

Divide Row2 by 1.60 or Multiply by 1/1.6
Add   -15•Row2 + Row1:Row1
Add   +16.1•Row2 + Row3:Row3

Column 6 still has a negative in the last row

column 6		last column	
2.8125	...	2850	2850/2.8125 ≈ 1013.3 ← row
0.8125	...	1250	1250/0.8125 ≈ 1538.5
-1.31875	...	20125	

29. Continued

   Pivot on 2.8125 in Row 1, Column 6

   Divide Row1 by 2.8125 or Multiply by 1/2.8125
   Add $-0.8125 \cdot$ Row1 + Row2:Row2
   Add $+1.31875 \cdot$ Row1 + Row3:Row3

There are no more negative numbers in the last row

Determine and check the best possible solution

   $(x_1, x_2, x_3, x_4, x_5, x_6, s_1, s_2)$
   $= (0, 0, 0, 0, 427, 1{,}013, 0, 0)$
   with $z = 21{,}461.33$

Check:
C_1: $50x_1 + 50x_2 + 35x_3 + 30x_4 + 15x_5 + 15x_6 + s_1$
   $= 0 + 0 + 0 + 0 + 15(427) + 15(1{,}013) + 0$
   $= 6{,}405 + 15{,}195 = 21{,}600$

C_2: $1.05x_1 + 0.95x_2 + 2.00x_3 + 1.55x_4 + 1.60x_5 + 1.30x_6 + s_2$
   $= 0 + 0 + 0 + 0 + 1.60(427) + 1.30(1{,}013) + 0$
   $= 683.20 + 1{,}316.9 = 2{,}000.10 \approx 2{,}000$

$z = 0.60x_1 + 0.70x_2 + 2.50x_3 + 2.00x_4 + 16.10x_5 + 14.40x_6$
   $= 0 + 0 + 0 + 0 + 16.10(427) + 14.40(1{,}013) + 0$
   $= 6{,}874.7 + 14{,}587.2 = 21{,}461.90 \approx 21{,}461.33$

Interpret the final solution

   Each week the Five Friends Bakery should make 427 dozen
   Blueberry muffins, 1,013 dozen apple-cinnamon muffins and
   nothing else. If they do that, they will use all the hours
   available and all the money available and make a weekly
   profit of approximately $21,461.33.

Exercise 8.3

33. **Step 1:** *Model the problem*

Independent variables:
x_1 = liters of House White wine produced this season
x_2 = liters of Premium White wine produced this season
x_3 = liters of Sauvignon Blanc wine produced this season
x_4 = liters of Chardonnay wine produced this season

	French Colombard	Chenin Blanc	Sauvignon Blanc	Chardonnay	Profit
House White	40%	40%	20%		$1.00
Premium White	25%		75%		$1.50
Sauvignon Blanc			100%		$2.25
Chardonnay		10%		90%	$3.00
Pounds Available	30,000	25,000	20,000	20,000	
Liters Possible*	15,000	12,500	10,000	10,000	

*2 pounds of grapes = 1 liter of wine

Constraints:
C_1: French Colombard grapes
French Colombard (FC) available $\leq$ 15,000
(FC in House) + (FC in Premium) $\leq$ 15,000
(40% of House) + (25% of Premium) $\leq$ 15,000
$0.40x_1 + 0.25x_2 \leq 15,000$

C_2: Chenin Blanc (CB) grapes
Chenin Blanc (CB) available $\leq$ 12,500
(CB in House) + (CB in Chardonnay) $\leq$ 12,500
(40% of House) + (10% of Chardonnay) $\leq$ 12,500
$0.40x_1 + 0.10x_4 \leq 12,500$

33. Continued

    C_3:   Sauvignon Blanc grapes

$$\text{Sauvignon Blanc (SB) available} \leq 10,000$$
$$\text{(SB in House)} + \text{(SB in Premium)} + \text{(SB in Sauvignon)} \leq 10,000$$
$$\text{(20\% of House)} + \text{(75\% of Premium)} + \text{(100\% of Sauvi)} \leq 10,000$$
$$0.20x_1 + 0.75x_2 + 1x_3 \leq 10,000$$

    C_4:   Chardonnay grapes

$$\text{Chardonnay (C) available} \leq 10,000$$
$$\text{(C in Chardonnay)} \leq 10,000$$
$$\text{(90\% of Chardonnay)} \leq 10,000$$
$$0.90x_4 \leq 10,000$$

Objective Function:

$$z = \text{(House profit)} + \text{(Premium profit)}$$
$$+ \text{(Sauvignon profit)} + \text{(Chardonnay profit)}$$
$$= (\$1 \text{ per liter})(\text{liters of House White})$$
$$+ (\$1.50 \text{ per liter})(\text{liters of Premium})$$
$$+ (\$2.25 \text{ per liter})(\text{liters of Sauvignon Blanc})$$
$$+ (\$3.00 \text{ per liter})(\text{liters of Chardonnay})$$

$$z = 1x_1 + 1.5x_2 + 2.25x_3 + 3.00x_4$$

Step 2: *Find the constraint equations*

    C_1: $0.40x_1 + 0.25x_2 + 0x_3 + 0x_4 + s_1 = 15,000$
             where s_1 = unused French Colombard
    C_2: $0.40x_1 + 0x_2 + 0x_3 + 0.10x_4 + s_2 = 12,500$
             where s_2 = unused Chenin Blanc
    C_3: $0.20x_1 + 0.75x_2 + 1.00x_3 + 0x_4 + s_3 = 10,000$
             where s_3 = unused Sauvignon Blanc
    C_4: $0x_1 + 0x_2 + 0x_3 + 0.90x_4 + s_4 = 10,000$
             where s_4 = unused Chardonnay

Step 3: *Rewrite the objective function equation*

$$z = 1.00x_1 + 1.50x_2 + 2.25x_3 + 3.00x_4 \text{ becomes}$$
$$-1.00x_1 - 1.50x_2 - 2.25x_3 - 3.00x_4 + 0s_1 + 0s_2 + 0s_3 + 0s_4 + 1z = 0$$

33. Continued

Step 4: *Find the first simplex matrix*

x_1	x_2	x_3	x_4	s_1	s_2	s_3	s_4	z		
0.40	0.25	0	0	1	0	0	0	0	15,000	Undef.
0.40	0	0	0.10	0	1	0	0	0	12,500	125,000
0.20	0.75	1.00	0	0	0	1	0	0	10,000	Undef
0	0	0	**0.90**	0	0	0	1	0	10,000	11,111
−1.00	−1.50	−2.25	−3.00	0	0	0	0	1	0	

$$\uparrow$$
pivot column

Step 5: *Find the possible solution*

$$(x_1, x_2, x_3, x_4, s_1, s_2, s_3, s_4)$$
$$= (0, 0, 0, 0, 15,000, 12,500, 10,000, 10,000)$$
with z = 0

Step 6: *Pivot to find the best possible solution*

Use the computer or calculator to find the best possible solution

Enter the matrix from above

Matrix dimensions are 5 × 10

Pivot on 0.90 in row 4, column 4

Divide Row4 by 0.90 or Multiply by 1/0.90
Add   −0.10•Row4 + Row2:Row2
Add   +3•Row4 + Row5:Row5

The most negative number in the last row is −2.25 in Column 3

column 3		last column		
0	...	15,000	Undefined	
0	...	11,388.88	Undefined	
1	...	10,000	10,000	
0	...	11,111.11	Undefined	
−2.25	...	33,333.33		

Pivot on 1 in Row 3, Column 3

Add   +2.25•Row3 + Row5:Row5

The most negative number in the last row is −0.55 in Column 1

33. Continued

column 1   last column

$$\begin{bmatrix} 0.4 & \dots & 15,000 & \bigg| & 37,500 \\ \mathbf{0.4} & \dots & 11,388.88 & \bigg| & 28,472.22 \\ 0.2 & \dots & 10,000 & \bigg| & 50,000 \\ 0 & \dots & 11,111.11 & \bigg| & \text{Undefined} \\ -0.55 & \dots & 55,833.33 & \bigg| & \end{bmatrix}$$

Pivot on 0.4 in Row 2, Column 1

Divide Row2 by 0.4 or Multiply by 1/0.4
Add   -0.4•Row2 + Row1:Row1
Add   +0.2•Row2 + Row3:Row3
Add   +0.55•Row2 + Row5:Row5

Last row is all positive, so we do not need to pivot again.

Step 7: *Determine and check the best possible solution*

x_1	x_2	x_3	x_4	s_1	s_2	s_3	s_4	z	
0	0.25	0	0	1	-1	0	0.11	0	3,611
1	0	0	0	0	2.5	0	-0.3	0	28,472
0	0.75	1	0	0	-0.5	1	0.06	0	4,306
0	0	0	1	0	0	0	1.11	0	11,111
0	0.19	0	0	0	1.38	2.25	3.18	1	71,493

$(x_1, x_2, x_3, x_4, s_1, s_2, s_3, s_4)$
= (28,472, 0, 4,306, 11,111, 3611, 0, 0, 0)
with z = \$71,493

Check:

C_1: $0.40x_1 + 0.25x_2 + 0x_3 + 0x_4 + s_1$
= 0.40(28,472) + 0 + 0 + 0 + (3,611)
= 11,388.8 + 3,611 = 14,999.8 ≈ 15,000

C_2: $0.40x_1 + 0x_2 + 0x_3 + 0.10x_4 + s_2$
= 0.40(28,472) + 0 + 0 + 0.10(11,111) + 0
= 11,388.8 + 1,111.1 = 12,499.9 ≈ 12,500

C_3: $0.20x_1 + 0.75x_2 + 1.00x_3 + 0x_4 + s_3$
= 0.20(28,472) + 0 + 1.00(4,306) + 0 + 0
= 5,694.4 + 4,306 = 10,000.4 ≈ 10,000

C_4: $0x_1 + 0x_2 + 0x_3 + 0.90x_4 + s_4 = 10,000$
= 0 + 0 + 0 + 0.90(11,111) + 0 = 9,999.9 ≈ 12,500

z = $1.00x_1 + 1.50x_2 + 2.25x_3 + 3.00x_4$
= 1.00(28,472) + 0 + 2.25(4,306) + 3.00(11,111)
= 28,472 + 9,688.50 + 33,333
= 71,493.5 ≈ 71,493

33. Continued

   Step 8: *Interpret the final solution*

   J & M Winery should make 28,742 liters of House White, no Premium White, 4,306 liters of Sauvignon Blanc, and 11,111 liters of Chardonnay to maximize its profit. The estimated profit would be $71,493. There would be 3,611(2 pounds/liter) = 7,222 pounds of unused French Colombard grapes.

Chapter 8 Review

1. **Solve the inequality for y:**

   $$4x - 5y > 7 \quad \rightarrow \quad -5y > -4x + 7 \quad \rightarrow \quad y < \frac{4}{5}x - \frac{7}{5}$$

   Graph the line: $y = (4/5)x + (7/5)$ is a dashed line (= is not part of the inequality) with slope (4/5) and y-intercept $-(7/5)$. Start at $(0,-7/5)$ on the y-axis and from that point rise 4 (move four units up) and run 5 (move five units to the right).

   $$\text{slope} = m = \frac{\text{rise}}{\text{run}} = \frac{4}{5} = \frac{4}{5} = \frac{\text{four units up}}{\text{five units right}}$$

   Shade in one side of the line: The graph of the inequality is the set of all points below the line (values of y decrease if we move down).

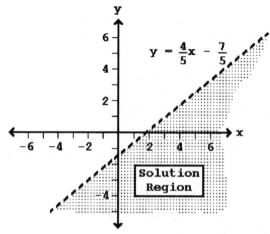

286

5.

Original Inequality	Slope-Intercept Form	Associated Equation	Graph of the Inequality
$8x - 4y < 10$	$-4y < -8x + 10$ $y > 2x - \dfrac{5}{2}$	$y = 2x - \dfrac{5}{2}$	all points above the line with slope 2 and y-intercept $-(5/2)$
$3x + 5y \geq 7$	$5y \geq -3x + 7$ $y \geq -\dfrac{5}{5}x + \dfrac{7}{5}$	$y = -\dfrac{3}{5}x + \dfrac{7}{5}$	all points on or above the line with slope $-(3/5)$ and y-intercept $7/5$

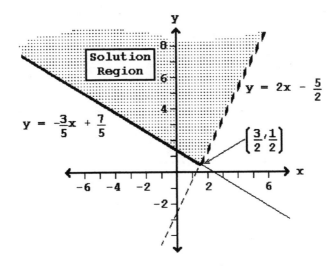

There is one corner point where the lines intersect. Use the elimination method.

$$y = 2x - \frac{5}{2} \quad \rightarrow \quad 10y = 20x - 25$$

$$y = -\frac{3}{5}x + \frac{7}{5} \quad \rightarrow \quad \underline{-10y = 6x - 14}$$

$$0 = 26x - 39$$

$$-26x = -39 \qquad \rightarrow \quad x = \frac{-39}{-26} = \frac{3}{2}$$

$$y = 2x - \frac{5}{2} = 2\left(\frac{3}{2}\right) - \frac{5}{2} = \frac{6}{2} - \frac{5}{2} = \frac{1}{2}$$

Thus, the corner point is $\left(\dfrac{3}{2}, \dfrac{1}{2}\right)$.

9.

Original Inequality	Associated Equation	Graph of the Inequality
Slope-Intercept Form		
$x - y \geq 7$ $-y \geq -x + 7 \quad \rightarrow$ $y \leq x - 7$	$y = x - 7$	all points on or below the line with slope 1 and y-intercept -7
$5x + 3y \leq 9$ $3y \leq -5x + 9 \quad \rightarrow$ $y \leq -\frac{5}{3}x + 3$	$y = -\frac{5}{3}x + 3$	all points on or below the line with slope -(5/3) and y-intercept 3
$x \geq 0$ (not applicable)	$x = 0$	all points on or to the right of the y-axis
$y \geq 0$ $y \geq 0$	$y = 0$	all points on or above the x-axis

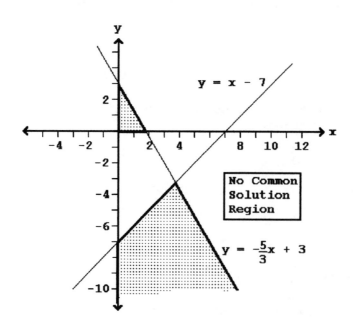

9. Continued

   There are no corner points because there is not a common solution region for the four inequalities.

13. **Step 1:** *Model the problem*

   Independent variables:
   x_1 = number of Model 110 produced
   x_2 = number of Model 330 produced

   Constraints:
   C_1: tweeters
   tweeters available $\leq$ 90
   (tweeters in Model 110) + (tweeters in Model 330) $\leq$ 90
   (1 tweeter)(Model 110) + (2 tweeters)(Model 330) $\leq$ 90
   $1x_1 + 2x_2 \leq 90$

   C_2: mid-range speakers
   mid-range speakers available $\leq$ 60
   (mid-range in Model 110) + (mid-range in Model 330) $\leq$ 60
   (1 mid-range)(Model 110) + (1 mid-range)(Model 330) $\leq$ 60
   $1x_1 + 1x_2 \leq 60$

   C_3: woofers
   woofers available $\leq$ 44
   (woofers in Model 110) + (woofers in Model 330) $\leq$ 44
   (0 woofers)(Model 110) + (1 woofer)(Model 330) $\leq$ 44
   $0x_1 + 1x_2 \leq 44$

   Objective Function:
   z = (Model 110 income) + (Model 330 income)
   = ($200)(Model 110) + ($350)(Model 330)

   $z = 200x_1 + 350x_2$

Step 2: *Find the constraint equations*

   C_1: $1x_1 + 2x_2 + s_1 = 90$ where s_1 = unused tweeters
   C_2: $1x_1 + 1x_2 + s_2 = 60$ where s_2 = unused mid-range
   C_3: $0x_1 + 1x_2 + s_3 = 44$ where s_3 = unused woofers

Step 3: *Rewrite the objective function equation*

   $z = 200x_1 + 350x_2$ becomes
   $-200x_1 - 350x_2 + 0s_1 + 0s_2 + 0s_3 + 1z = 0$

Exercise 8.3

13. Continued

Step 4: *Find the first simplex matrix*

$$
\begin{array}{cccccc}
x_1 & x_2 & s_1 & s_2 & s_3 & z \\
\end{array}
$$

$$
\left[
\begin{array}{cccccc|c}
1 & 2 & 1 & 0 & 0 & 0 & 90 \\
1 & 1 & 0 & 1 & 0 & 0 & 60 \\
0 & 1 & 0 & 0 & 1 & 0 & 44 \\
-200 & -350 & 0 & 0 & 0 & 1 & 0 \\
\end{array}
\right]
$$

Step 5: *Find the possible solution*

$$(x_1, x_2, s_1, s_2, s_3) = (0, 0, 90, 60, 44) \text{ with } z = 0$$

Step 6: *Pivot to find a better possible solution*

$$
\begin{array}{cccccc}
x_1 & x_2 & s_1 & s_2 & s_3 & z \\
\end{array}
$$

$$
\left[
\begin{array}{cccccc|c}
1 & 2 & 1 & 0 & 0 & 0 & 90 \\
1 & 1 & 0 & 1 & 0 & 0 & 60 \\
0 & 1 & 0 & 0 & 1 & 0 & 44 \\
-200 & -350 & 0 & 0 & 0 & 1 & 0 \\
\end{array}
\right]
\begin{array}{l}
90/2 = 45 \\
60/1 = 60 \\
44/1 = 44 \leftarrow \text{pivot row} \\
\\
\end{array}
$$

↑
pivot column

Pivot on 1 in row 3, column 2

$$
\begin{array}{r}
-2R3 + R1:R1 \\
-R3 + R2:R2 \\
\\
350R3 + R4:R4 \\
\end{array}
\left[
\begin{array}{cccccc|r}
\mathbf{1} & 0 & 1 & 0 & -2 & 0 & 2 \\
1 & 0 & 0 & 1 & -1 & 0 & 16 \\
0 & 1 & 0 & 0 & 1 & 0 & 44 \\
-200 & 0 & 0 & 0 & 350 & 1 & 15,400 \\
\end{array}
\right]
\begin{array}{l}
2 \leftarrow \text{pivot row} \\
16 \\
\\
\\
\end{array}
$$

↑
pivot column

Last row is not all positive, so pivot again.

Pivot on 1 in row 1, column 1

$$
\begin{array}{r}
\\
-R1 + R2:R2 \\
\\
200R1 + R4:R4 \\
\end{array}
\left[
\begin{array}{cccccc|r}
1 & 0 & 1 & 0 & -2 & 0 & 2 \\
0 & 0 & -1 & 1 & \mathbf{1} & 0 & 14 \\
0 & 1 & 0 & 0 & 1 & 0 & 44 \\
0 & 0 & 200 & 0 & -50 & 1 & 15,800 \\
\end{array}
\right]
\begin{array}{l}
-1 \\
14 \leftarrow \text{pivot row} \\
44 \\
\\
\end{array}
$$

↑
pivot column

13. Continued

   Last row is not all positive, so pivot again.

   Pivot on 1 in row 2, column 5

$$
\begin{array}{r}
2R2 + R1:R1 \\
\\
-R2 + R3:R3 \\
50R2 + R4:R4
\end{array}
\left[
\begin{array}{ccccccc}
1 & 0 & -1 & 2 & 0 & 0 & 30 \\
0 & 0 & -1 & 1 & 1 & 0 & 14 \\
0 & 1 & 1 & -1 & 0 & 0 & 30 \\
0 & 0 & 150 & 50 & 0 & 1 & 16,500
\end{array}
\right]
$$

Last row is all positive, so we do not need to pivot again.

Step 7: *Determine the possible solution*

   $(x_1, x_2, s_1, s_2, s_3) = (30, 30, 0, 0, 14)$ with $z = 16,500$

   Check:

   C_1: $1x_1 + 2x_2 + s_1$
   $1(30) + 2(30) + 0 = 30 + 60 = 90$

   C_2: $1x_1 + 1x_2 + s_2$
   $1(30) + 1(30) + 0 = 30 + 30 = 60$

   C_3: $0x_1 + 1x_2 + s_3$
   $0(30) + 1(30) + 14 = 0 + 30 + 14 = 44$

   $z = 200x_1 + 350x_2$
   $= 200(30) + 350(30) = 6,000 + 10,500 = 16,500$

Step 8: *Determine if this maximizes the objective function*

   The last row contains no negative entries, so we have maximized the income.

Step 9: *Interpret the final solution*

   Mowson Audio Co. should make 30 model 110 speaker assemblies and 30 model 330 assemblies to maximize its income. The maximum income will be $16,500. They would use all the tweeters and all the mid-range speakers, but would have 14 woofers left over.

17. Omitted.

9 Exponential and Logarithmic Functions

Exercise 9.0A

1. $v = \log_2 8$ can be rewritten as $2^v = 8$.

   $2^v = 8 = 2^3$, so $v = 3$

5. $\log_5 u = -2$ can be rewritten as $5^{-2} = u$.

   $u = 5^{-2} = \dfrac{1}{25}$

9. $\log_b 9 = 2$ can be rewritten as $b^2 = 9$.

   $b^2 = 9 = 3^2$, so $b = 3$ (base can only be positive)

13. $K = \log_b H$ can be rewritten as $b^K = H$.

17. $b^T = S$ can be rewritten as $T = \log_b S$

21. a) $e^{1.2} = 3.320116923$

    b) $10^{1.2} = 15.84893192$

25. a) $\dfrac{1}{e^{1.5}} = e^{-1.5} = 0.2231301601$

    b) $\dfrac{1}{10^{1.5}} = 10^{-1.5} = 0.0316227766$

29. a) $\dfrac{e^{4.7}}{2e^{5.1}} = \dfrac{e^{4.7-5.1}}{2} = \dfrac{e^{-0.4}}{2} = \dfrac{0.67032005}{2} = 0.335160023$

    b) $\dfrac{10^{4.7}}{2(10)^{5.1}} = \dfrac{10^{4.7-5.1}}{2} = \dfrac{10^{-0.4}}{2} = \dfrac{0.39810717}{2} = 0.1990535853$

33. a) $\ln(0.58) = -0.5447271754$

    b) $\log(0.58) = -0.2365720064$

37. a) $\ln(4e^{0.02}) = \ln[(4)(1.02020134)] = \ln(4.08080536)$
       $= 1.406294361$

    b) $\log[4(10)^{0.02}] = \log[(4)(1.04712855)] = \log(4.18851419)$
       $= 0.622059913$

41. a) $e^{2\ln 5} = e^{2(1.60943791)} = e^{3.21887583} = 25$

    b) $10^{2\log 5} = 10^{2(0.69897000)} = 10^{1.39794001} = 25$

45. a) $\ln(10^{2.47}) = \ln(295.120923) = 5.68738518$

    b) $\log(10^{2.47}) = \log(295.120923) = 2.47$

49. Omitted.

Exercise 9.0B

1. $\log(10^{7.5x}) = 7.5x$

5. $10^{\log(3x + 1)} = 3x + 1$

9. $\log\left(\dfrac{x}{5}\right) = \log x - \log 5$ Division-Becomes-Subtraction

13. $\log(1.0625^{x}) = x\log(1.0625)$ Exponent-Becomes-Multiplier

17. $\ln\left(\dfrac{2x}{3}\right) = \ln(2x) - \ln 3$ Division-Becomes-Subtraction
    $= \ln 2 + \ln x - \ln 3$ Multiplication-Becomes-Addition

21. $\ln(5x) + \ln 2 = \ln[(5x)(2)]$ Reverse Mult-Becomes-Addition
    $= \ln(10x)$ Simplify

Exercise 9.0B

25. $2\ln(3x) - \ln 9 = \ln(3x)^2 - \ln 9$ Exponent-Becomes-Multiplier

$$= \ln\frac{(3x)^2}{9} \qquad \text{Division-Becomes-Subtraction}$$

$$= \ln\frac{9x^2}{9} = \ln(x^2) \qquad \text{Simplify}$$

29. a) $e^x = 0.25$

     $\ln(e^x) = \ln(0.25)$ Taking ln of each side

        $x = \ln(0.25)$ Inverse Property

        $x = -1.386294361$ Calculator

     Check: $e^x = e^{-1.386294361} = 0.25$

     b) $10^x = 0.25$

     $\log(10^x) = \log(0.25)$ Taking log of each side

        $x = \log(0.25)$ Inverse Property

        $x = -0.602059991$ Calculator

     Check: $10^x = 10^{-0.602059991} = 0.25$

33. a) $1000e^{0.009x} = 1500$

$$e^{0.009x} = \frac{1500}{1000} = 1.5 \qquad \text{Isolate the exponential}$$

$$\ln(e^{0.009x}) = \ln(1.5) \qquad \text{Taking ln of each side}$$

$$0.009x = \ln(1.5) \qquad \text{Inverse Property}$$

$$x = \frac{\ln(1.5)}{0.009}$$

$$x = \frac{0.40546511}{0.009} = 45.05167868 \qquad \text{Calculator}$$

$$\text{\textit{Check:} } 1000e^{0.009x} = 1000e^{0.009(45.05167868)}$$

$$= 1000e^{0.40546511}$$

$$= 1000(1.5)$$

$$= 1500$$

33. Continued

b) $1000(10)^{0.009x} = 1500$

$$10^{0.009x} = \frac{1500}{1000} = 1.5 \quad \text{Isolate the exponential}$$

$$\log(10^{0.009x}) = \log(1.5) \quad \text{Taking log of each side}$$

$$0.009x = \log(1.5) \quad \text{Inverse Property}$$

$$x = \frac{\log(1.5)}{0.009}$$

$$x = \frac{0.17609126}{0.009} = 19.56569545 \quad \text{Calculator}$$

Check: $1000(10)^{0.009x} = 1000(10)^{0.009(19.56569545)}$

$$= 1000(10)^{0.17609126}$$

$$= 1000(1.5)$$

$$= 1500$$

37. a) $50e^{-0.0016x} = 40$

$$e^{-0.0016x} = \frac{40}{50} = 0.8 \quad \text{Isolate the exponential}$$

$$\ln(e^{-0.0016x}) = \ln(0.8) \quad \text{Taking ln of each side}$$

$$-0.0016x = \ln(0.8) \quad \text{Inverse Property}$$

$$x = \frac{\ln(0.8)}{-0.0016}$$

$$x = \frac{-0.22314355}{-0.0016} = 139.4647196 \quad \text{Calculator}$$

Check: $50e^{-0.0016x} = 50e^{-0.0016(139.4647196)}$

$$= 50e^{-0.22314355}$$

$$= 50(0.8) = 40$$

b) $50(10)^{-0.0016x} = 40$

$$10^{-0.0016x} = \frac{40}{50} = 0.8 \quad \text{Isolate the exponential}$$

$$\log(10^{-0.0016x}) = \log(0.8) \quad \text{Taking log of each side}$$

$$-0.0016x = \log(0.8) \quad \text{Inverse Property}$$

37.b)   Continued

$$x = \frac{\log(0.8)}{-0.0016}$$

$$x = \frac{-0.09691001}{-0.0016} = 60.56875813 \quad \text{Calculator}$$

Check: $50(10)^{-0.0016x} = 50(10)^{-0.0016(60.56875813)}$

$$= 50(10)^{-0.09691001}$$

$$= 50(0.8) = 40$$

41.   a) $\log x = 1.85$

$10^{\log x} = 10^{1.85}$ Exponentiate each side

$x = 10^{1.85}$ Inverse Property

$x = 70.79457844$ Calculator

Check: $\log x = \log(70.79457844) = 1.85$

b) $\ln x = 1.85$

$e^{\ln x} = e^{1.85}$ Exponentiate each side

$x = e^{1.85}$ Inverse Property

$x = 6.359819523$ Calculator

Check: $\ln x = \ln(6.359819523) = 1.85$

45.   a) $\log x = 1.8 + \log(3.6)$

$\log x - \log(3.6) = 1.8$ Collect log terms

$\log\left(\frac{x}{3.6}\right) = 1.8$ Division-Becomes-Subtraction

$10^{\log(x/3.6)} = 10^{1.8}$ Exponentiate each side

$\frac{x}{3.6} = 10^{1.8}$ Inverse Property

$x = 3.6(10)^{1.8}$

$x = 3.6(63.09573445)$ Calculator

$= 227.144644$

45. Continued

Check:
$$\log x = 1.8 + \log(3.6)$$
$$\log(227.144644) = 1.8 + \log(3.6)$$
$$2.3563025 = 1.8 + 0.5563025$$
$$2.3563025 = 2.3563025$$

b)
$$\ln x = 1.8 - \ln(3.6)$$

$\ln x + \ln(3.6) = 1.8$	Collect ln terms
$\ln[(x)(3.6)] = 1.8$	Multiplication–Becomes–Addition
$e^{\ln(3.6x)} = e^{1.8}$	Exponentiate each side
$3.6x = e^{1.8}$	Inverse Property

$$x = \frac{e^{1.8}}{3.6}$$

$$x = \frac{6.04964746}{3.6} = 1.680457629 \qquad \text{Calculator}$$

Check:
$$\ln x = 1.8 - \log(3.6)$$
$$\ln(1.680457629) = 1.8 - \ln(3.6)$$
$$0.51906615 = 1.8 - 1.28093385$$
$$0.51906615 = 0.51906615$$

49. Show that $\ln(A/B) = \ln A - \ln B$

Let $u = \ln A$ which can be rewritten as $e^u = A$

Let $v = \ln B$ which can be rewritten as $e^v = B$

$\ln\left(\dfrac{A}{B}\right) = \ln\left(\dfrac{e^u}{e^v}\right)$	Substitution
$= \ln(e^{u-v})$	Exponent law
$= u - v$	Inverse property
$= \ln A - \ln B$	Substitution

53. Omitted.

Exercise 9.1

---

1. $p = 30e^{0.0198026273t}$, $t = 2001 - 1997 = 4$ years

    $p = 30e^{0.0198026273(4)} = 32.47296480 \approx 32.473$ thousand or 32,473

    The population of Anytown is predicted to be 34,461 in 2001.

5. a) In 1980, $t = 0$ and $p = 8115$ thousand so ordered pair is (0, 8115).
        In 1990, $t = 10$ and $p = 8240$ thousand so ordered pair is (10, 8240).

    b) $\Delta t = 10 - 0 = 10$ years

    c) $\Delta p = 8240 - 8115 = 125$ thousand

    d) $\dfrac{\Delta p}{\Delta t} = \dfrac{125}{10} = 12.5$ thousand people per year

    e) $\dfrac{\Delta p / \Delta t}{p} = \dfrac{12.5}{8115} = 0.001540357 \approx 0.15\%$ per year

9. a) In 1980, time 0, $(t,p) = (0, 8115)$
        In 1990, $(t,p) = (10, 8240)$

        Substitute the first ordered pair (0, 8115)

        $$p = ae^{bt}$$

        $$8115 = ae^{b(0)}$$

        $$8115 = ae^{0}$$

        $$8115 = a(1)$$

        $$a = 8115 \qquad \text{(the initial value of p)}$$

        Rewrite model to $p = 8115e^{bt}$

9.a) Continued

Substitute the second ordered pair (10, 8240)

$$8240 = 8115e^{b(10)}$$

$$8240 = 8115e^{10b}$$

$$\frac{8240}{8115} = e^{10b}$$

$$1.01540357 = e^{10b}$$

$$\ln(1.01540357) = \ln(e^{10b})$$

$$10b = \ln(1.01540357)$$

$$b = \frac{\ln(1.01540357)}{10} = \frac{0.015286143}{10}$$

$$= 0.0015286143$$

The model is $p = 8115e^{0.0015286143t}$, t in years.

b) In 1991, t = 1991 − 1980 = 11 years

$$p = 8115e^{0.0015286143(11)} = 8252.6054 \approx 8253 \text{ thousand}$$

In 1991, the predicted population was 8253 thousand people.

c) In 2000, t = 2000 − 1980 = 20 years

$$p = 8115e^{0.0015286143(20)} = 8366.9254 \approx 8367 \text{ thousand}$$

In 2000, the predicted population is 8367 thousand people.

d) Double the 1980 population = 2(8115)

$$2(8115) = 8115e^{0.0015286143t}$$

$$\frac{2(8,115)}{8,115} = e^{0.0015286143t}$$

$$2 = e^{0.0015286143t}$$

$$\ln 2 = \ln(e^{0.0015286143t})$$

Exercise 9.1

9.d)   Continued

$$0.0015286143t = \ln 2$$

$$t = \frac{\ln 2}{0.0015286143} = 453.4481 \approx 453 \text{ years}$$

In 2433 AD (1980 + 453 years) the population will have doubled.

13.   a)   Time 0, $(t,p) = (0, 2510)$
Three days later, $(t,p) = (3, 5380)$

Substitute the first ordered pair $(0, 2510)$

$$p = ae^{bt}$$

$$2510 = ae^{b(0)}$$

$$2510 = ae^{0}$$

$$2510 = a(1)$$

$$a = 2510 \qquad \text{(the initial value of p)}$$

Rewrite model to $p = 2510a^{bt}$

Substitute the second ordered pair $(3, 5380)$

$$5380 = 2510e^{b(3)}$$

$$5380 = 2510e^{3b}$$

$$\frac{5380}{2510} = e^{3b}$$

$$\ln\left(\frac{5380}{2510}\right) = \ln(e^{3b})$$

$$3b = \ln\left(\frac{5380}{2510}\right)$$

$$b = 0.2541352$$

The model is $p = 2510e^{0.2541352t}$, t in days.

13. **Continued**

    b)   In one week, t = 7 days

    $$p = 2510e^{0.2541352(7)} = 14,868.27 \approx 14,870$$

    In one week the predicted population is 14,870 fruit flies (rounded to the nearest 10).

    c)   Double the population = 2(2510)

    $$2(2510) = 2510e^{0.2541352t}$$

    $$\frac{2(2510)}{2510} = e^{0.2541352t}$$

    $$2 = e^{0.2541352t}$$

    $$\ln 2 = \ln(e^{0.2541352t})$$

    $$0.2541352t = \ln 2$$

    $$t = \frac{\ln 2}{0.2541352} = 2.72747 \approx 2.7 \text{ days}$$

    The population will double in approximately 2.7 days.

17. a)   In 1992, time 0, (t,A) = (0, 85.2), A in quadrillion Btu's
          In 1993, (t,A) = (1, 86.9), A in quadrillion Btu's

    Substitute the first ordered pair (0, 85.2)

    $$A = ae^{bt}$$

    Since a is the initial value of A, a = 85.2.

    Rewrite model to $A = 85.2e^{bt}$

    Substitute the second ordered pair (1, 86.9)

    $$86.9 = 85.2e^{b(1)}$$

    $$86.9 = 85.2e^{b}$$

    $$\frac{86.9}{85.2} = e^{b}$$

Exercise 9.1

17.a)   Continued

$$\ln\left(\frac{86.9}{85.2}\right) = \ln(e^b)$$

$$b = \ln\left(\frac{86.9}{85.2}\right) = 0.0197565984$$

The model is $A = 85.2e^{0.0197565984t}$, t in years.

b)   In 1994, t = 1994 - 1992 = 2 years

$$A = 85.2e^{0.0197565984(2)} = 88.63392 \approx 88.6$$

In 1994, the predicted energy consumed was 88.6 quadrillion Btu's.

c)   Omitted.

21. a)   In 1790, time 0, (t,p) = (0, 3,929,214)
In 1800, (t,p) = (10, 5,308,483)

Substitute the first ordered pair (0, 3,929,214)
$$p = ae^{bt}$$

Since a is the initial value of p, a = 3,929,214.

Rewrite model to $p = 3,929,214e^{bt}$

Substitute the second ordered pair (10, 5,308,483)

$$5,308,483 = 3,929,214e^{b(10)}$$

$$5,308,483 = 3,929,214e^{10b}$$

$$\frac{5,308,483}{3,929,214} = e^{10b}$$

$$\ln\left(\frac{5,308,483}{3,929,214}\right) = e^{10b}$$

$$10b = \ln\left(\frac{5,308,483}{3,929,214}\right)$$

$$b = 0.03008667$$

The model is $p = 3,929,214e^{0.03008667t}$, t in years.

21. Continued

    b)   The 1810 prediction would be more accurate.

    c)   In 1810, t = 1810 − 1790 = 20 years

$$p = 3,929,214e^{0.03008667(20)} = 7,171,915.747 \approx 7,171,916$$

    In 1810, the predicted population was 7,171,916 people.

    In 1990, t = 1990 − 1790 = 200 years

$$p = 3,929,214e^{0.03008667(200)} = 1,612,874,720$$

    In 1990, the predicted population was 1,612,874,720 people.

25. In 1980, time 0, (t,p) = (0, 4.478 billion)
    In 1991, (t,p) = (11, 5.423 billion)

    The model is $p = 4.478e^{0.0174066t}$

    Double the population = 2(4.478)

$$2(4.478) = 4.478e^{0.0174066t}$$

$$2 = e^{0.0174066t}$$

$$\ln 2 = \ln(e^{0.0174066t})$$

$$0.0174066t = \ln 2$$

$$t = \frac{\ln 2}{0.0174066} = 39.8209 \approx 40 \text{ years}$$

    The article claims the world population will double in 39
    years. We calculate it will double in approximately 40
    years which is a difference of one year or an error of 1/39
    = 0.025641 ≈ 2.6%.

29. Omitted.

Exercise 9.1

33. From Exercise 7:   In 1980, time 0, $(t,p) = (0, 18,713)$
                        In 1990, $(t,p) = (10, 19,342)$

$FV = 19,342$, $P = 18,713$, $n = 10$

Substitute in $FV = P(1 + i)^n$ and solve for $(1 + i)$

$$19,342 = 18,713(1 + i)^{10}$$

$$\frac{19,342}{18,713} = (1 + i)^{10} \qquad \text{Dividing by } 18,713$$

$$\ln\left(\frac{19,342}{18,713}\right) = \ln(1 + i)^{10} \qquad \text{Taking ln of each side}$$

$$\ln\left(\frac{19,342}{18,713}\right) = 10\ln(1 + i) \qquad \text{Exponent-Becomes-Multiplier}$$

$$\frac{1}{10}\left(\ln\left(\frac{19,342}{18,713}\right)\right) = \ln(1 + i) \qquad \text{Dividing by } 10$$

$$0.0033060428 = \ln(1 + i)$$

$$e^{0.0033060428} = 1 + i \qquad \text{Apply the definition}$$

$$1.003311514 = 1 + i$$

$$FV = 18,713(1.003311514)^t$$

37.
$$2P = P\left(1 + \frac{0.05}{365}\right)^n$$

$$2 = (1 + 0.000369863)^n = (1.000369863)^n$$

$$\ln 2 = n\ln(1.000369863)$$

$$n = \frac{\ln 2}{\ln(1.000369863)} = 5060.3 \text{ days} \approx 13.86 \text{ years}$$

Exercise 9.2

1. $Q = 20e^{-0.086643397t}$, $t = 3$ weeks $= 21$ days

$$Q = 20e^{-0.086643397(21)} = 3.24209893 \approx 3.2$$

After 3 weeks approximately 3.2 grams of iodine-131 are left.

5.   Half-life of silicon-31 is 2.6 hours

a)   Ordered pairs (t,Q) are (0,a) and (2.6,a/2)

Omit substituting the first pair since we are not using a specific value.

Model is $Q = ae^{bt}$

Substitute the second ordered pair (2.6,a/2)

$$\frac{a}{2} = ae^{b(2.6)}$$

$$\frac{1}{2} = e^{2.6b} \qquad \text{dividing by a}$$

$$0.5 = e^{2.6b}$$

$$\ln(0.5) = \ln(e^{2.6b})$$

$$2.6b = \ln(0.5)$$

$$b = \frac{\ln(0.5)}{2.6} = -0.266595069$$

The model is $Q = ae^{-0.266595069t}$, t in hours.

b)   In one hour, t = 1, a = 50 milligrams

$$Q = 50e^{-0.266595069(1)} = 38.299159$$

In one hour the predicted amount of silicon-31 is 38.3 milligrams.

c)   In one day, t = 24 hours, a = 50 milligrams

$$Q = 50e^{-0.266595069(24)} = 0.0832207$$

In one day the predicted amount of silicon-31 is 0.08 milligrams.

d)   Ordered pairs at one hour: (0,50) and (1,38.299159)

$$\frac{\Delta Q}{\Delta t} = \frac{38.299159 - 50}{1 - 0} = \frac{-11.70084}{1}$$

$$= -11.70084 \approx -11.7 \text{ milligrams per hour}$$

e) $$\frac{\Delta Q/\Delta t}{Q} = \frac{-11.70084}{50} = -0.2340168 = -23.4\% \text{ per hour}$$

Exercise 9.2

5. Continued

   f) Ordered pairs at one day: (0,50) and (24,0.0832207)

$$\frac{\Delta Q}{\Delta t} = \frac{0.0832207 - 50}{24 - 0} = \frac{-49.916779}{24}$$

$$= -2.079866 \approx -2.1 \text{ milligrams per hour}$$

   g) $\frac{\Delta Q / \Delta t}{Q} = \frac{-2.079866}{50} = -0.0415973 \approx -4.2\% \text{ per hour}$

   h) Radioactive substances decay faster when there is more substance present. A larger quantity is lost during the first part of the day than during the last part of the day and when the relative changes are computed the relative decay rate will be greater for shorter periods of time.

9. Half-life of plutonium-241 is 13 years
Ordered pairs (t,Q) are (0,a) and (13,a/2)

Omit substituting the first pair since we are not using a specific value.

$$\text{Model is } Q = ae^{bt}$$

Substitute second ordered pair (13,a/2)

$$\frac{a}{2} = ae^{b(13)}$$

$$\frac{1}{2} = e^{13b} \qquad \text{dividing by a}$$

$$0.5 = e^{13b}$$

$$\ln(0.5) = \ln(e^{13b})$$

$$13b = \ln(0.5)$$

$$b = \frac{\ln(0.5)}{13} = -0.053319014$$

The model is $Q = ae^{-0.053319014t}$, t in years.

$Q = 100$ grams, $a = 500$ grams, find t

$$100 = 500e^{-0.053319014t}$$

$$\frac{100}{500} = e^{-0.053319014t}$$

$$0.2 = e^{-0.053319014t}$$

$$\ln(0.2) = \ln(e^{-0.053319014t})$$

9.   Continued

$$-0.053319014t = \ln(0.2)$$

$$t = \frac{\ln(0.2)}{-0.053319014} = 30.185065 \approx 30.2 \text{ years}$$

It would take 30.2 years for plutonium-241 to decay from 500 grams to 100 grams.

13.   Half-life of plutonium-239 is 24,400 years
      Ordered pairs (t,Q) are $(0,a)$ and $(24,400,a/2)$

      Omit substituting the first pair since we are not using a specific value.

      Model is $Q = ae^{bt}$

      Substitute second ordered pair $(24,400,a/2)$

$$\frac{a}{2} = ae^{b(24,400)}$$

$$\frac{1}{2} = e^{24,400b} \qquad \text{dividing by a}$$

$$0.5 = e^{24,400b}$$

$$\ln(0.5) = \ln(e^{24,400b})$$

$$24,400b = \ln(0.5)$$

$$b = \frac{\ln(0.5)}{24,400} = -0.00002840767$$

The model is $Q = ae^{-0.00002840767t}$, t in years.

If 90% of its radioactivity is lost, 10% remains.

$Q = 0.10$, $a = 1.00$, find t

$$0.1 = 1.0e^{-0.00002840767t}$$

$$0.1 = e^{-0.00002840767t}$$

$$\ln(0.1) = \ln(e^{-0.00002840767t})$$

$$-0.00002840767t = \ln(0.1)$$

$$t = \frac{\ln(0.1)}{-0.00002840767}$$

$$= 81,055.0455 \approx 81,055 \text{ years}$$

It would take 81,055 years for plutonium-239 to lose 90% of its radioactivity.

Exercise 9.2

17. The radiocarbon dating model is $Q = ae^{-0.000120968t}$, t in years.
    t = 5250

$$Q = ae^{-0.000120968(5250)} = a(0.529892) \approx 0.53a$$

You would expect to find 53% of the original amount of carbon-14.

21. The radiocarbon dating model is $Q = ae^{-0.000120968t}$, t in years.
    There is 84% of the expected carbon-14 remaining, Q = 0.84a.

$$0.84a = ae^{-0.000120968t}$$

$$0.84 = e^{-0.000120968t}$$

$$\ln(0.84) = \ln(e^{-0.000120968t})$$

$$-0.000120968t = \ln(0.84)$$

$$t = \frac{\ln(0.84)}{-0.000120968t} = 1441.3183 \approx 1441 \text{ years}$$

The age of the roof material and therefore the age of the Mayan codex would be approximately 1441 years.

25. The radiocarbon dating model is $Q = ae^{-0.000120968t}$, t in years.
    There is 70% of the expected carbon-14 remaining, Q = 0.70a.

$$0.70a = ae^{-0.000120968t}$$

$$0.70 = e^{-0.000120968t}$$

$$\ln(0.70) = \ln(e^{-0.000120968t})$$

$$-0.000120968t = \ln(0.70)$$

$$t = \frac{\ln(0.70)}{-0.000120968} = 2948.5066 \approx 2949 \text{ years}$$

The age of the parchment would be approximately 2949 years.

29. The radiocarbon dating model is $Q = ae^{-0.000120968t}$, t in years.
    t = 5730

$$Q = ae^{-0.000120968(5730)} = a(0.5000) = 0.5a$$

You would expect to find 50% of the original amount of carbon-14. (5730 is the half-life of carbon-14.)

33. Omitted.

37. Omitted.

Exercise 9.3

1. A = 3.9 × 10⁴ μm at 100 km from the epicenter, find M.

$$M = \log A - \log A_0$$

$$= \log(3.9 \times 10^4) - (-3.0) \qquad \text{From Figure 9.24}$$

$$= 4.591064607 + 3.0$$

$$= 7.59106 \approx 7.6$$

The magnitude of the earthquake was 7.6 on the Richter scale.

5. $M_1 = 8.3 = $ San Francisco (1906)
$M_2 = 7.1 = $ San Francisco (1989)

a) Use the magnitude comparison formula

$$M_1 - M_2 = \log\left(\frac{A_1}{A_2}\right)$$

$$8.3 - 7.1 = \log\left(\frac{A_1}{A_2}\right)$$

$$1.2 = \log\left(\frac{A_1}{A_2}\right)$$

$$10^{1.2} = 10^{\log(A_1/A_2)}$$

$$10^{1.2} = \frac{A_1}{A_2}$$

$$A_1 = 10^{1.2}A_2 = 15.84893A_2 \approx 16A_2$$

The 1906 earthquake caused about 16 times as much earth movement as the 1989 earthquake.

Exercise 9.3

5.   Continued

    b)   Use the energy formula

$$\log E \approx 11.8 + 1.45M$$

    For 1906

$$\log E_1 \approx 11.8 + 1.45(8.3) = 23.835$$

$$10^{\log E_1} \approx 10^{23.835}$$

$$E_1 \approx 10^{23.835}$$

    For 1989

$$\log E_2 \approx 11.8 + 1.45(7.1) = 22.095$$

$$10^{\log E_2} \approx 10^{22.095}$$

$$E_2 \approx 10^{22.095}$$

Comparing energies: $\dfrac{E_1}{E_2} \approx \dfrac{10^{23.835}}{10^{22.095}} = 10^{1.74} = 54.954087$

$$E_1 \approx 55E_2$$

The 1906 earthquake released 55 times as much energy as the 1989 earthquake.

9. $M_1 = 7.7 = $ Iran
    $M_2 = 6.2 = $ Turkey

    a)   Use the magnitude comparison formula

$$M_1 - M_2 = \log\left(\frac{A_1}{A_2}\right)$$

$$7.7 - 6.2 = \log\left(\frac{A_1}{A_2}\right)$$

$$1.5 = \log\left(\frac{A_1}{A_2}\right)$$

9.a)   Continued

$$10^{1.5} = 10^{\log(A_1/A_2)} = \frac{A_1}{A_2}$$

$$A_1 = 10^{1.5}A_2 = 31.62278A_2 \approx 32A_2$$

The Iranian earthquake caused about 32 times as much earth movement as the earthquake in Turkey.

b)   Use the energy formula

$$\log E \approx 11.8 + 1.45M$$

For Iran

$$\log E_1 \approx 11.8 + 1.45(7.7) = 22.965$$

$$10^{\log E_1} \approx 10^{22.965}$$

$$E_1 \approx 10^{22.965}$$

For Turkey

$$\log E_2 \approx 11.8 + 1.45(6.2) = 20.790$$

$$10^{\log E_2} \approx 10^{20.790}$$

$$E_2 \approx 10^{20.790}$$

Comparing energies: $\quad \dfrac{E_1}{E_2} \approx \dfrac{10^{22.965}}{10^{20.790}} = 10^{2.175} = 149.6236$

$$E_1 \approx 150E_2$$

The Iranian earthquake released 150 times as much energy as the earthquake in Turkey.

13. $M_1 = 7.1$ = final estimate
      $M_2 = 7.0$ = original estimate

a)   Use the magnitude comparison formula

$$M_1 - M_2 = \log\left(\frac{A_1}{A_2}\right)$$

$$7.1 - 7.0 = \log\left(\frac{A_1}{A_2}\right)$$

13.a) Continued

$$0.1 = \log\left(\frac{A_1}{A_2}\right)$$

$$10^{0.1} = 10^{\log(A_1/A_2)} = \frac{A_1}{A_2}$$

$$A_1 = 10^{0.1}A_2 = 1.258925A_2 \approx 1.26A_2$$

The change in magnitude readings from 7.0 to 7.1 corresponds to a 26% increase in earth movement.

b) Use the energy formula

$$\log E \approx 11.8 + 1.45M$$

<u>For final estimate</u>
$$\log E_2 \approx 11.8 + 1.45(7.1) = 22.095$$

$$10^{\log E_2} \approx 10^{22.095}$$

$$E_2 \approx 10^{22.095}$$

<u>For original estimate</u>
$$\log E_2 \approx 11.8 + 1.45(7.0) = 21.950$$

$$10^{\log E_2} \approx 10^{21.950}$$

$$E_2 \approx 10^{21.950}$$

Comparing energies: $\dfrac{E_1}{E_2} \approx \dfrac{10^{22.095}}{10^{21.950}} = 10^{0.145} = 1.396368$

$$E_1 \approx 1.396368E_2 \approx 1.40E_2$$

The change in magnitude from 7.0 to 7.1 increased the energy release by 40%.

17. Use the decibel rating definition with $I_0 \approx 10^{-16}$ watts/cm^2.
$I = 10^{-9}$ watts/cm^2

$$D = 10\log\left(\frac{I}{I_0}\right)$$

$$= 10\log\left(\frac{10^{-9}}{10^{-16}}\right) = 10\log\left(10^{-9 - (-16)}\right) = 10\log(10^7)$$

$$= 10(7) = 70$$

The decibel rating of the television is 70 dB.

21. Use the decibel gain formula

$I_1 = 10^{-13}$ watts/cm^2 , $I_2 = 10^{-14}$ watts/cm^2

$$D_1 - D_2 = 10\log\left(\frac{I_1}{I_2}\right)$$

$$= 10\log\left(\frac{10^{-13}}{10^{-14}}\right)$$

$$= 10\log\left(10^{-13 - (-14)}\right)$$

$$= 10\log(10^1) = 10(1) = 10$$

The decibel gain is 10 dB.

25. I_2 = sound intensity of single singer
I_1 = sound intensity with additional singers
D_1 = 81 dB, D_2 = 74 dB

$$D_1 - D_2 = 10\log\left(\frac{I_1}{I_2}\right)$$

$$81 - 74 = 10\log\left(\frac{I_1}{I_2}\right)$$

$$7 = 10\log\left(\frac{I_1}{I_2}\right)$$

$$0.7 = \log\left(\frac{I_1}{I_2}\right)$$

$$10^{0.7} = 10^{\log(I_1/I_2)}$$

25. Continued

$$\frac{I_1}{I_2} = 10^{0.7} = 5.0118723 \approx 5 \text{ total singers}$$

Four singers have joined the original singer for a total of 5 singers.

29. I_2 = sound intensity of single trumpet
    I_1 = sound intensity with additional trumpets
    $D_1 = 85.8 \text{ dB}, \ D_2 = 78 \text{ dB}$

$$D_1 - D_2 = 10\log\left(\frac{I_1}{I_2}\right)$$

$$85.8 - 78 = 10\log\left(\frac{I_1}{I_2}\right)$$

$$7.8 = 10\log\left(\frac{I_1}{I_2}\right)$$

$$0.78 = \log\left(\frac{I_1}{I_2}\right)$$

$$10^{0.78} = 10^{\log(I_1/I_2)}$$

$$\frac{I_1}{I_2} = 10^{0.78} = 6.02559586 \approx 6 \text{ trumpets}$$

There are five additional trumpets for a total of 6 trumpets.

33. $D_1 = 105 \text{ dB}$ = original volume
    $D_2 = 60 \text{ dB}$ = reduced volume

$$D_1 - D_2 = 10\log\left(\frac{I_1}{I_2}\right)$$

$$105 - 60 = 10\log\left(\frac{I_1}{I_2}\right)$$

$$45 = 10\log\left(\frac{I_1}{I_2}\right)$$

$$\frac{45}{10} = \log\left(\frac{I_1}{I_2}\right)$$

33. Continued

$$4.5 = \log\left(\frac{I_1}{I_2}\right)$$

$$10^{4.5} = 10^{\log(I_1/I_2)}$$

$$\frac{I_1}{I_2} = 10^{4.5} = 31,622.78 \approx 31,623$$

$$I_1 \approx 31,623I_2 \quad \text{or} \quad I_2 \approx \frac{1}{31,623} I_1 \approx 0.0000316 2I_1$$

The original volume was about 32,000 times the sound intensity of the reduced volume. Or, the reduced volume would be 0.003% as intense as the original volume.

37. Omitted.

Chapter 9 Review

1. $x = \log_3 81$ can be rewritten as $3^x = 81$.

$3^x = 81 = 3^4$, so $x = 4$

5. $\ln(e^x) = x$ and $e^{\ln x} = x$ are the inverse properties of the natural logarithm.

9. $\ln(A^n) = n(\ln A)$ is the Exponent-Becomes-Multiplier property of the natural logarithm.

13. $\log(x + 2)$ cannot be rewritten using the properties of logarithms.

17. $\log(5x) + \log x^2 - \log x = 12$

$\log[5x(x^2)] - \log x = 12$ Multiplication-Becomes-Addition

$\log\dfrac{5x^3}{x} = 12$ Division-Becomes-Subtraction

$\log(5x^2) = 12$ Simplify

$10^{\log(5x^2)} = 10^{12}$ Exponentiate each side

$5x^2 = 10^{12}$ Inverse Property

$x^2 = \dfrac{10^{12}}{5} = 2 \times 10^{11}$

$x = 447{,}213.6$ Taking square root

21. Closest: A = 25 μm at 20 km from the epicenter, $A_0 = -1.7$

$M = \log A - \log A_0$

$= \log(25) - (-1.7)$ From Figure 9.24

$= 1.397940 + 1.7$

$= 3.097940 \approx 3.1$

Second: A = 2 μm at 60 km from the epicenter, $A_0 = -2.8$

$M = \log A - \log A_0$

$= \log(2) - (-2.8)$ From Figure 9.24

$= 0.30103 + 2.8$

$= 3.10103 \approx 3.1$

The magnitude of the earthquake was 3.1 on the Richter scale.

25. I_2 = sound intensity of single trumpet
      I_1 = sound intensity with additional trumpets
      D_1 = 84 dB, D_2 = 78 dB

$$D_1 - D_2 = 10 \log \left(\frac{I_1}{I_2} \right)$$

$$84 - 78 = 10 \log \left(\frac{I_1}{I_2} \right)$$

$$6 = 10 \log \left(\frac{I_1}{I_2} \right)$$

$$0.6 = \log \left(\frac{I_1}{I_2} \right)$$

$$10^{0.6} = 10^{\log(I_1/I_2)}$$

$$\frac{I_1}{I_2} = 10^{0.6} = 3.98107 \approx 4 \text{ trumpets}$$

There are three additional trumpets making a total of four trumpets.

29.   Omitted.

10 Calculus

Exercise 10.0

1. Similar triangles: 4 corresponds to 2, 5 corresponds to a,
    and 7 corresponds to b

    <u>Solving for a</u>   <u>Solving for b</u>

    $\dfrac{4}{2} = \dfrac{5}{a}$ $\dfrac{4}{2} = \dfrac{7}{b}$

    $4a = 5(2)$ $4b = 7(2)$

    $4a = 10$ $4b = 14$

    $a = \dfrac{10}{4} = \dfrac{5}{2}$ $b = \dfrac{14}{4} = \dfrac{7}{2}$

5. Similar triangles: 6 corresponds to 9, 8 corresponds to x,
    and 10 corresponds to (10 + y)

    <u>Solving for x</u>   <u>Solving for y</u>

    $\dfrac{6}{9} = \dfrac{8}{x}$ $\dfrac{6}{9} = \dfrac{10}{10 + y}$

    $6x = 8(9)$ $6(10 + y) = 9(10)$

    $6x = 72$ $60 + 6y = 90$

    $x = \dfrac{72}{6} = 12$ $6y = 90 - 60 = 30$

                                $y = \dfrac{30}{6} = 5$

9. Student-teacher ratio of 15 to 1

    a) $\dfrac{15 \text{ students}}{1 \text{ teacher}}$ or $\dfrac{1 \text{ teacher}}{15 \text{ students}}$

    b) $\dfrac{15 \text{ students}}{1 \text{ teacher}} = \dfrac{5430 \text{ students}}{x \text{ teachers}}$

             $15x = 1(5430)$

             $x = \dfrac{5430}{15} = 362$ faculty members

13. Odometer: beginning = 5.4, ending = 332.5, gallons = 13.3

   a) Change in Distance = (ending − beginning) odometer
   = 332.5 − 5.4 = 327.1 miles

   $$\frac{\text{change in distance}}{\text{gallons consumed}} = \frac{327.1 \text{ mi}}{13.3 \text{ gal}} = 24.59398$$

   $$\approx 24.6 \text{ mpg (miles per gallon)}$$

   b) $\dfrac{24.59398 \text{ miles}}{1 \text{ gallon}} = \dfrac{x}{15 \text{ gallons}}$

   x = 24.59398(15) = 368.9097 ≈ 369 miles

   c) miles = mpg(gallon) = 24.6(15) = 369 miles

17. See Figure 10.3. b = 756 feet, S = 342 feet,
   p = 6 feet, s = 9 feet

   Similar triangles: h corresponds to p, $\left(\frac{1}{2}b + S\right)$ corresponds to s

   $$\frac{h}{p} = \frac{\frac{1}{2}b + S}{s}$$

   $$\frac{h}{6} = \frac{\frac{1}{2}(756) + 342}{9} = \frac{378 + 342}{9} = \frac{720}{9}$$

   9h = 6(720) = 4320

   $$h = \frac{4320}{9} = 480 \text{ feet}$$

21. $y = -x^2 - 4x + 3$

   a) Substituting 0, 1, and 2 for x yields the following points:

x	$y = -x^2 - 4x + 3$	y	(x, y)
0	$-(0)^2 - 4(0) + 3 = -0 - 0 + 3$	3	(0, 3)
1	$-(1)^2 - 4(1) + 3 = -1 - 4 + 3$	−2	(1, −2)
2	$-(2)^2 - 4(2) + 3 = -4 - 8 + 3$	−9	(2, −9)

21.a)   Continued

> Look at the slope of the secant lines to determine which direction the vertex lies:
>
> Points $(0,3)$ and $(1,-2)$: $m = \dfrac{\Delta y}{\Delta x} = \dfrac{-2 - 3}{1 - 0} = -5$
>
> Points $(1,-2)$ and $(2,-9)$: $m = \dfrac{\Delta y}{\Delta x} = \dfrac{-9 - (-2)}{2 - 1} = -7$
>
> The vertex lies to the left of $x = 0$ since -5 is closer to zero than -7.

x	$y = -x^2 - 4x + 3$	y	(x, y)
-1	$-(-1)^2 - 4(-1) + 3 = -1 + 4 + 3$	6	(-1, 6)
-2	$-(-2)^2 - 4(-2) + 3 = -4 + 8 + 3$	7	(-2, 7)
-3	$-(-3)^2 - 4(-3) + 3 = -9 + 12 + 3$	6	(-3, 6)
-4	$-(-4)^2 - 4(-4) + 3 = -16 + 16 + 3$	3	(-4, 3)
-5	$-(-5)^2 - 4(-5) + 3 = -25 + 20 + 3$	-2	(-5, -2)

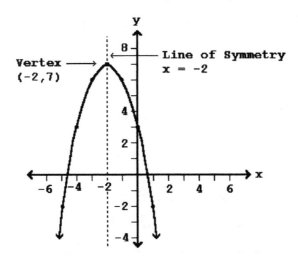

b)   The y-values are the same at $x = -1$ and $x = -3$, so the equation of the line of symmetry is $x = -2$.

c)   The vertex is the point $(-2, 7)$.

25. $y = 8x - x^2 - 14$ or, in standard form, $y = -x^2 + 8x - 14$

a) Substituting 0, 1, and 2 for x yields the following points:

x	$y = -x^2 + 8x - 14$	y	(x,y)
0	$(0)^2 + 8(0) - 14 = 0 + 0 - 14$	-14	(0,-14)
1	$-(1)^2 + 8(1) - 14 = -1 + 8 - 14$	-7	(1,-7)
2	$-(2)^2 + 8(2) - 14 = -4 + 16 - 14$	-2	(2,-2)

Look at the slope of the secant lines to determine which direction the vertex lies:

Points (0,-14) and (1,-7): $m = \dfrac{\Delta y}{\Delta x} = \dfrac{-7 - (-14)}{1 - 0} = 7$

Points (1,-7) and (2,-2): $m = \dfrac{\Delta y}{\Delta x} = \dfrac{-2 - (-7)}{2 - 1} = 5$

The vertex lies to the right of x = 2 since 5 is closer to zero than 7.

x	$y = -x^2 + 8x - 14$	y	(x,y)
3	$-(3)^2 + 8(3) - 14 = -9 + 24 - 14$	1	(3,1)
4	$-(4)^2 + 8(4) - 14 = -16 + 32 - 14$	2	(4,2)
5	$-(5)^2 + 8(5) - 14 = -25 + 40 - 14$	1	(5,1)

Exercise 10.0

25.a)   Continued

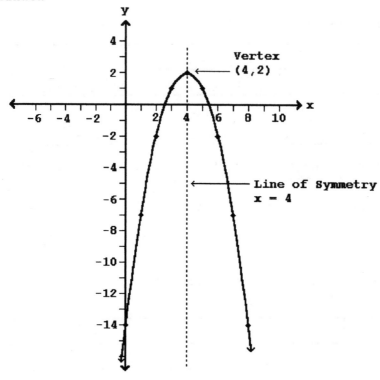

b)   The y-values are the same at x = 3 and x = 5, so the
     equation of the line of symmetry is x = 4.

c)   The vertex is the point (4,2).

29.   y = 2 - 3x² or, in standard form, y = -3x² + 2; x = 1, x = 3

a) x = 1:   y = -3(1)² + 2 = -3 + 2 = -1
   x = 3:   y = -3(3)² + 2 = -3(9) + 2 = -27 + 2 = -25

   Using points (1,-1) and (3,-25), the slope is

$$m = \frac{\Delta y}{\Delta x} = \frac{-25 - (-1)}{3 - 1} = \frac{-24}{2} = -12$$

29. Continued

b) Substituting for x yields the following points:

x	$y = 2 - 3x^2$	y	(x,y)
-2	$2 - 3(-2)^2 = 2 - 3(4) = 2 - 12$	-10	(-2,-10)
-1	$2 - 3(-1)^2 = 2 - 3(1) = 2 - 3$	-1	(-1,-1)
0	$2 - 3(0)^2 = 2 - 3(0) = 2 - 0$	2	(0,2)
1	$2 - 3(1)^2 = 2 - 3(1) = 2 - 3$	-1	(1,-1)
2	$2 - 3(2)^2 = 2 - 3(4) = 2 - 12$	-10	(2,-10)
3	$2 - 3(3)^2 = 2 - 3(9) = 2 - 27$	-25	(3,-25)

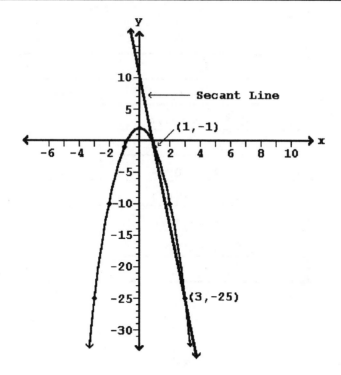

33. $f(x) = 8x - 11$
    $f(4)$ means "substitute 4 for x in function f"
    $$f(4) = 8(4) - 11$$
    $$= 32 - 11 = 21$$

323

Exercise 10.0

37. f(x) = 8x – 11
     f(x + 3) means "substitute (x + 3) for x in function f"
         f(x + 3) = 8(x + 3) – 11
                  = 8x + 24 – 11 = 8x + 13

41. f(x) = 8x – 11
     f(x + Δx) means "substitute (x + Δx) for x in function f"
         f(x + Δx) = 8(x + Δx) – 11
                   = 8x + 8Δx – 11

Exercise 10.1

1. $y = \frac{1}{4}x^2 = (x^2)/4$

    a)

x	$y = (x^2)/4$	y	(x,y)
–3	$(-3)^2/4 = (9)/4$	9/4	(–3,9/4)
–2	$(-2)^2/4 = (4)/4$	1	(–2,1)
–1	$(-1)^2/4 = (1)/4$	1/4	(–1,1/4)
0	$(0)^2/4 = (0)/4$	0	(0,0)
1	$(1)^2/4 = (1)/4$	1/4	(1,1/4)
2	$(2)^2/4 = (4)/4$	1	(2,1)
3	$(3)^2/4 = (9)/4$	9/4	(3,9/4)
4	$(4)^2/4 = (16)/4$	4	(4,4)

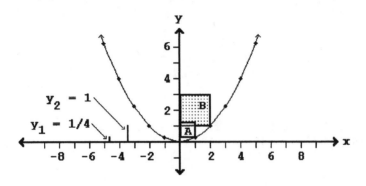

1.   Continued

b)   At point $(1, 1/4)$ the line segment is of length $y_1 = 1/4$. The base of the square is of length $x_1 = 1$, so the area of the square is $1^2 = 1$. See square A above.

At point $(2, 1)$, the line segment is of length $y_2 = 1$. The base of the square is of length $x_2 = 2$, so the area of the square is $2^2 = 4$. See square B above.

c)
$$\frac{\text{length of line segment 1}}{\text{length of line segment 2}} = \frac{y_1}{y_2} = \frac{\frac{1}{4}}{1} = \frac{1}{4}$$

$$\frac{\text{area of square A}}{\text{area of square B}} = \frac{1}{4}$$

d)   See computations in 1.a) above for graph.

At point $(3, 9/4)$ the line segment is of length $y_3 = 9/4$. The base of the square is of length $x_3 = 3$, so the area of the square is $3^2 = 9$. See square C below.

At point $(4, 4)$, the line segment is of length $y_4 = 4$. The base of the square is of length $x_4 = 4$, so the area of the square is $4^2 = 16$. See square D below.

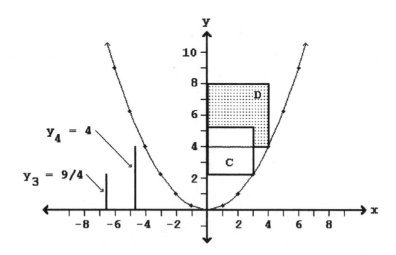

Exercise 10.1

1. Continued

    e) $\dfrac{\text{length of line segment 3}}{\text{length of line segment 4}} = \dfrac{y_3}{y_4} = \dfrac{\frac{9}{4}}{4} = \left(\dfrac{9}{4}\right)\left(\dfrac{1}{4}\right) = \dfrac{9}{16}$

    $\dfrac{\text{area of square C}}{\text{area of square D}} = \dfrac{9}{16}$

5. $y = x^2 + 1$

    a)

x	$y = x^2 + 1$	y	(x,y)
-3	$(-3)^2 + 1 = 9 + 1$	10	(-3,10)
-2	$(-2)^2 + 1 = 4 + 1$	5	(-2,5)
-1	$(-1)^2 + 1 = 1 + 1$	2	(-1,2)
0	$(0)^2 + 1 = 0 + 1$	1	(0,1)
1	$(1)^2 + 1 = 1 + 1$	2	(1,2)
2	$(2)^2 + 1 = 4 + 1$	5	(2,5)
3	$(3)^2 + 1 = 9 + 1$	10	(3,10)

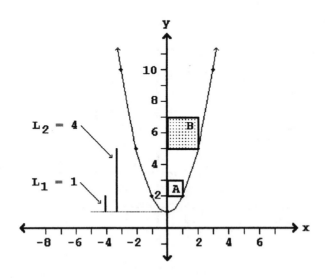

5.   Continued

b)   At point (1,2) the line segment is of length L_1 = 1. The
      base of the square is of length x_1 = 1, so the area of the
      square is 1^2 = 1. See square A above.

      At point (2,5), the line segment is of length L_2 = 4.   The
      base of the square is of length x_2 = 2, so the area of the
      square is 2^2 = 4. See square B above.

c)
      $$\frac{\text{length of line segment 1}}{\text{length of line segment 2}} = \frac{L_1}{L_2} = \frac{1}{4}$$

      $$\frac{\text{area of square A}}{\text{area of square B}} = \frac{1}{4}$$

9.   a)   A square and 12 roots are equal to 45 units. To solve this
            problem take 1/2 the roots which would give 6. Add the
            square of this number (36) to 45 which gives 81. Take the
            square root of 81 which is 9 and subtract 1/2 of the roots
            (or 6). Hence the root is 3 and the square is 9.

      b)   **al-Khowarizmi's Geometric Justification:** Construct a square
            with unknown roots. Construct a rectangle with the same
            unknown root as a side and the other side of length 12.
            Divide the rectangle into 4 parts each with the unknown
            length and the other length equal to 12 divided by 4 which
            is 3.

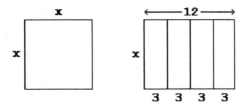

            Each strip is attached to one of the four sides of the
            square. In order to make a complete large square, four
            small squares with sides of length 3 are added to the
            figure.

9.b) Continued

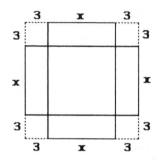

The area of one small square is 9. The resulting area is a
square with area of 45 plus 4(9) = 36 which is 81. Hence
each root of the large square is 9. But each side of the
large square is the unknown root plus 6, therefore the
unknown root is 3 and the area of the square is 9.

c) **Modern notation**

Solve $x^2 + 12x = 45$ for x^2.

$\frac{1}{2}(12) = 6$	Take half of the roots
$6^2 = 36$	Square result
$x^2 + 12x + 36 = 45 + 36$	Add result to both sides
$x^2 + 12x + 36 = 81$	Simplify
$(x + 6)^2 = 9^2$	Factor perfect squares
$x + 6 = 9$	Take positive square root only
$x = 3$	Solve equation
$x^2 = 9$	Square result

13. a) A square and 2 roots are equal to 80. To solve this problem
take 1/2 the roots which would be 1. Add the square of this
number to 80 which gives 81. Take the square root of 81,
which is 9, and subtract 1/2 of the roots or 1. Hence the
root is 8 and the square is 64.

b) **al-Khowarizmi's geometric justification:** Construct a square
with unknown side x. Construct a rectangle with the unknown
width x and length of 2. Divide the rectangle into 4 equal
rectangles each x wide and 1/2 unit long.

13.a)   Continued

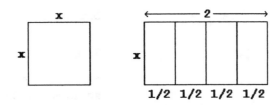

Each small rectangle is attached to a side of the initial square to make a figure with area of 80. Four small squares of area 1/4 square units are added to each corner of the figure making a complete larger square with area of 81.

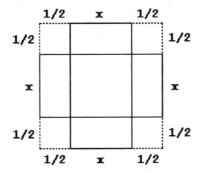

Hence, the larger square has a side of 9. But each side is the unknown plus 1, so the unknown root is 8 and the area of the square is 64.

c)   **Modern notation**

Solve $x^2 + 2x = 80$ for x^2.

$$\frac{1}{2}(2) = 1 \qquad \text{Take half of the roots}$$

$$1^2 = 1 \qquad \text{Square result}$$

$$x^2 + 2x + 1 = 80 + 1 \qquad \text{Add result to both sides}$$

$$x^2 + 2x + 1 = 81 \qquad \text{Simplify}$$

$$(x + 1)^2 = 9^2 \qquad \text{Factor perfect squares}$$

$$x + 1 = 9 \qquad \text{Take positive square root only}$$

$$x = 8 \qquad \text{Solve equation}$$

$$x^2 = 64 \qquad \text{Square result}$$

Exercise 10.1

17. Solve $ax^2 + bx = c$ for x.

$$\frac{ax^2}{a} + \frac{bx}{a} = \frac{c}{a}$$ Divide by a

$$x^2 + \frac{b}{a}x = \frac{c}{a}$$ Simplify

$$x^2 + \frac{b}{a}x + \frac{b^2}{4a^2} = \frac{c}{a} + \frac{b^2}{4a^2}$$ Add 1/2 of the x coefficient squared to both sides of the equation.

$$\left(x + \frac{b}{2a}\right)^2 = \frac{4ac + b^2}{4a^2}$$ Factor perfect square

$$x + \frac{b}{2a} = \sqrt{\frac{4ac + b^2}{4a^2}}$$ Take positive square roots

$$x + \frac{b}{2a} = \frac{\sqrt{4ac + b^2}}{2a}$$ Simplify

$$x = \frac{-b + \sqrt{4ac + b^2}}{2a}$$ Solve equation

21. Omitted.

25. Omitted.

29. Omitted.

33. Omitted.

37. Omitted.

41. Omitted.

45. Omitted.

Exercise 10.2

1. Time in motion until time B = AB = 10 seconds,
    Speed at time B = BC = 4 feet per second

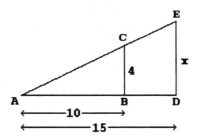

a) distance = area of triangle ABC = $\frac{1}{2}$bh

    d = $\frac{1}{2}$(10)(4) = 20 feet

b) time in motion until time D = AD = 15 seconds,
    speed at time D = x

    $\frac{\text{speed at time B}}{\text{speed at time D}} = \frac{\text{time in motion until time B}}{\text{time in motion until time D}}$

    $\frac{4}{x} = \frac{10}{15}$

    10x = 4(15) = 60

    x = $\frac{60}{10}$ = 6 feet per second

c) distance = area of triangle ADE = $\frac{1}{2}$bh

    d = $\frac{1}{2}$(15)(6) = 45 feet

d) average speed = $\frac{\text{distance traveled in 15 seconds}}{\text{15 seconds}}$

    average speed = $\frac{45}{15}$ = 3 feet per second

Exercise 10.2

5.

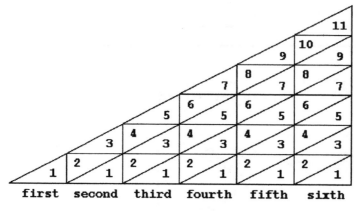

first   second   third   fourth   fifth   sixth

Interval of Time

Interval of Time	Distance Traveled during that Interval	Total Distance Traveled
first	1	1
second	3	1 + 3 = 4
third	5	1 + 3 + 5 = 9
fourth	7	1 + 3 + 5 + 7 = 16
fifth	9	1 + 3 + 5 + 7 + 9 = 25
sixth	11	1 + 3 + 5 + 7 + 9 + 11 = 36

9.   First Trip: distance = 40 feet, water = 7 ounces
      Second Trip: distance = 10 feet, water = x

$$\frac{\text{distance from first trip}}{\text{distance from second trip}} = \frac{(\text{water from first trip})^2}{(\text{water from second trip})^2}$$

$$\frac{40}{10} = \frac{7^2}{x^2}$$

$$\frac{40}{10} = \frac{49}{x^2}$$

$$40x^2 = 49(10)$$

$$x^2 = \frac{490}{40} = \frac{49}{4} \quad \text{so } x = \frac{7}{2} = 3.5 \text{ ounces}$$

13.   First Trip: distance = 1600 feet, time = 10 seconds
      Second Trip: distance = ?, time = 1 second

$$\frac{\text{distance from first trip}}{\text{distance from second trip}} = \frac{(\text{time of first trip})^2}{(\text{time of second trip})^2}$$

$$\frac{1600}{x} = \frac{10^2}{1^2}$$

$$\frac{1600}{x} = \frac{100}{1}$$

$100x = 1600, \quad \text{so } x = \frac{1600}{100} = 16 \text{ feet}$

17. $y = 2x^2$, P is at $(3, 18)$, so Q is at $(3 - e, 18 - a)$

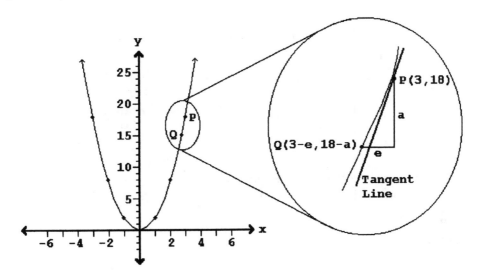

Solve for $\frac{a}{e}$

$$y = 2x^2$$
$$18 - a = 2(3 - e)^2$$
$$18 - a = 2(9 - 6e + e^2)$$
$$18 - a = 18 - 12e + 2e^2$$
$$2e^2 - 12e + a = 0 \qquad\qquad \text{eliminate powers of e and a}$$
$$-12e + a = 0$$
$$a = 12e$$

$$\frac{a}{e} = 12 = m = \text{slope of tangent line at } (3, 18)$$

Exercise 10.2

17. Continued

Equation of the tangent line:

$$y - y_1 = m(x - x_1) \quad \text{Solve for y where } (x_1, y_1) = (3, 18)$$
$$y - 18 = 12(x - 3)$$
$$y - 18 = 12x - 36$$
$$y = 12x - 18$$

21. $y = x^2 - 2x + 1$, P is at $(3, 4)$, so Q is at $(3 - e, 4 - a)$

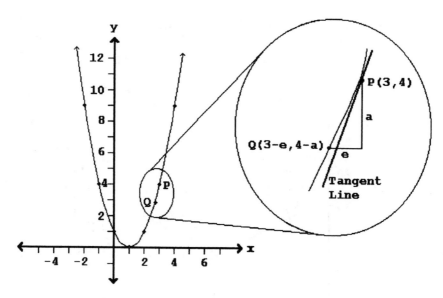

Solve for $\dfrac{a}{e}$

$$y = x^2 - 2x + 1$$
$$4 - a = (3 - e)^2 - 2(3 - e) + 1$$
$$4 - a = 9 - 6e + e^2 - 6 + 2e + 1$$
$$4 - a = 4 - 4e + e^2$$
$$e^2 - 4e + a = 0 \qquad \text{eliminate powers of e and a}$$
$$-4e + a = 0$$
$$a = 4e,$$

$\dfrac{a}{e} = 4 = m = $ slope of tangent line at $(3, 4)$

Solve for the tangent line:

$$y - y_1 = m(x - x_1) \quad \text{Solve for y where } (x_1, y_1) = (3, 4)$$
$$y - 4 = 4(x - 3)$$
$$y - 4 = 4x - 12$$
$$y = 4x - 8$$

25. $y = x^2 + 4x + 6$, P is at $(0,6)$, so Q is at $(0 - e, 6 - a)$

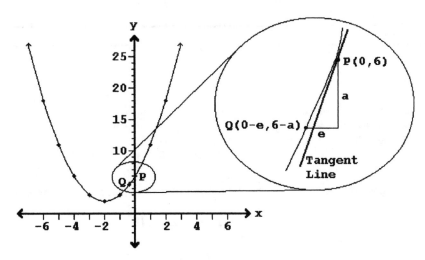

Solve for $\dfrac{a}{e}$

$$y = x^2 + 4x + 6$$
$$6 - a = (0 - e)^2 + 4(0 - e) + 6$$
$$6 - a = (-e)^2 + 4(-e) + 6$$
$$6 - a = e^2 - 4e + 6$$
$$- a = e^2 - 4e$$

$e^2 - 4e + a = 0$ \hspace{2cm} eliminate powers of e and a
$$-4e + a = 0$$
$$a = 4e,$$

$\dfrac{a}{e} = 4 = m = $ slope of tangent line at $(0,6)$

Solve for the tangent line:
$$y - y_1 = m(x - x_1) \hspace{1cm} \text{Solve for y where } (x_1, y_1) = (0,6)$$
$$y - 6 = 4(x - 0)$$
$$y - 6 = 4x - 0$$
$$y = 4x + 6$$

29. Omitted.

33. Omitted.

37. Omitted.

41. Omitted.

Exercise 10.3

1. $y = 3x^2$ at $(4, 48)$

a) **Newton's Method:**

P is at $x_1 = 4$, $y_1 = 3(4)^2 = 3(16) = 48$

Q is at $x_2 = 4 + o$, $y_2 = 3(4 + o)^2$
$$= 3(16 + 8o + o^2) = 48 + 24o + 3o^2$$

slope of PQ $= \dfrac{\Delta y}{\Delta x} = \dfrac{y_2 - y_1}{x_2 - x_1}$ where PQ is the secant line

$$= \frac{(48 + 24o + 3o^2) - 48}{(4 + o) - 4}$$

$$= \frac{24o + 3o^2}{o} = \frac{o(24 + 3o)}{o} = 24 + 3o$$

Let $o = 0$, then slope of tangent line $= 24 + 0 = 24$

b) **Cauchy's method:**

P is at $x_1 = 4$, $y_1 = 3(4)^2 = 48$

Q is at $x_2 = 4 + \Delta x$, $y_2 = 3(4 + \Delta x)^2 = 3(16 + 8\Delta x + \Delta x^2)$
$$= 48 + 24\Delta x + 3\Delta x^2$$

slope of PQ $= \dfrac{\Delta y}{\Delta x} = \dfrac{y_2 - y_1}{x_2 - x_1}$ where PQ is the secant line

$$= \frac{(48 + 24\Delta x + 3\Delta x^2) - 48}{(4 + \Delta x) - 4}$$

$$= \frac{24\Delta x + 3\Delta x^2}{\Delta x} = \frac{\Delta x(24 + 3\Delta x)}{\Delta x} = 24 + 3\Delta x$$

The closer Δx is to 0, the closer Q is to P, and the closer the slope of the secant line PQ is to the slope of the tangent line. If Δx is allowed to approach zero without reaching zero, then $24 + 3\Delta x$ will approach 24 without reaching 24. The slope of the tangent line at $(3, 48)$ is 24.

c) **Equation of tangent line:**

$y - y_1 = m(x - x_1)$ Solve for y where $(x_1, y_1) = (4, 48)$

$y - 48 = 24(x - 4)$

$y - 48 = 24x - 96$

$y = 24x - 48$

5. $y = 2x^2 - 5x + 1$ at $x = 7$

   a)   **Newton's Method:**

   P is at $x_1 = 7$, $y_1 = 2(7)^2 - 5(7) + 1 = 2(49) - 35 + 1 = 64$
   Q is at $x_2 = 7 + o$, $y_2 = 2(7 + o)^2 - 5(7 + o) + 1$
   $$= 2(49 + 14o + o^2) - 35 - 5o + 1$$
   $$= 98 + 28o + 2o^2 - 34 - 5o$$
   $$= 64 + 23o + 2o^2$$

   slope of PQ $= \dfrac{\Delta y}{\Delta x} = \dfrac{y_2 - y_1}{x_2 - x_1}$ where PQ is the secant line

   $$= \frac{(64 + 23o + 2o^2) - 64}{(7 + o) - 7}$$

   $$= \frac{23o + 2o^2}{o} = \frac{o(23 + 2o)}{o} = 23 + 2o$$

   Let $o = 0$, then slope of tangent line $= 23 + 0 = 23$

   b)   **Cauchy's method:**

   P is at $x_1 = 7$, $y_1 = 64$ (See Exercise 5.a)
   Q is at $x_2 = 7 + \Delta x$, $y_2 = 2(7 + \Delta x)^2 - 5(7 + \Delta x) + 1$
   $$= 2(49 + 14\Delta x + \Delta x^2) - 35 - 5\Delta x + 1$$
   $$= 98 + 28\Delta x + 2\Delta x^2 - 34 - 5\Delta x$$
   $$= 64 + 23\Delta x + 2\Delta x^2$$

   slope of PQ $= \dfrac{\Delta y}{\Delta x} = \dfrac{y_2 - y_1}{x_2 - x_1}$ where PQ is the secant line

   $$= \frac{(64 + 23\Delta x + 2\Delta x^2) - 64}{(7 + \Delta x) - 7}$$

   $$= \frac{23\Delta x + 2\Delta x^2}{\Delta x} = \frac{\Delta x(23 + 2\Delta x)}{\Delta x} = 23 + 2\Delta x$$

   The closer Δx is to 0, the closer Q is to P, and the closer
   the slope of the secant line PQ is to the slope of the
   tangent line.   If Δx is allowed to approach zero without
   reaching zero, then $23 + 2\Delta x$ will approach 23 without
   reaching 23.   The slope of the tangent line at $(7,64)$ is 23.

   c)   **Equation of tangent line:**

   $y - y_1 = m(x - x_1)$ Solve for y where $(x_1, y_1) = (7, 64)$

   $y - 64 = 23(x - 7)$
   $y - 64 = 23x - 161$
   $ y = 23x - 97$

Exercise 10.3

9. $y = x^3 - x^2$ at $x = 1$

a) **Newton's Method:**

P is at $x_1 = 1$, $y_1 = (1)^3 - (1)^2 = (1) - (1) = 0$

Q is at $x_2 = 1 + o$, $y_2 = (1 + o)^3 - (1 + o)^2$

$$= (1 + o)(1 + 2o + o^2) - (1 + 2o + o^2)$$
$$= (1 + 3o + 3o^2 + o^3) - 1 - 2o - o^2$$
$$= 0 + o + 2o^2 + o^3$$

slope of PQ $= \dfrac{\Delta y}{\Delta x} = \dfrac{y_2 - y_1}{x_2 - x_1}$ where PQ is the secant line

$$= \frac{(0 + o + 2o^2 + o^3) - 0}{(1 + o) - 1}$$

$$= \frac{o + 2o^2 + o^3}{o} = \frac{o(1 + 2o + o^2)}{o}$$

$$= 1 + 2o + o^2$$

Let $o = 0$, then slope of tangent line $= 1 + 0 + 0 = 1$

b) **Cauchy's Method:**

P is at $x_1 = 1$, $y_1 = 0$ (See Exercise 9.a)

Q is at $x_2 = 1 + \Delta x$, $y_2 = (1 + \Delta x)^3 - (1 + \Delta x)^2$

$$y_2 = (1 + 3\Delta x + 3\Delta x^2 + \Delta x^3) - (1 + 2\Delta x + \Delta x^2)$$
$$= 1 + 3\Delta x + 3\Delta x^2 + \Delta^3 - 1 - 2\Delta x - \Delta x^2$$
$$= 0 + \Delta x + 2\Delta x^2 + \Delta x^3$$

slope of PQ $= \dfrac{\Delta y}{\Delta x} = \dfrac{y_2 - y_1}{x_2 - x_1}$ where PQ is the secant line

$$= \frac{(0 + \Delta x + 2\Delta x^2 + \Delta x^3) - 0}{(1 + \Delta x) - 1}$$

$$= \frac{\Delta x + 2\Delta x^2 + \Delta x^3}{\Delta x} = \frac{\Delta x(1 + 2\Delta x + \Delta x^2)}{\Delta x}$$

$$= 1 + 2\Delta x + \Delta x^2$$

The closer Δx is to 0, the closer Q is to P, and the closer the slope of the secant line PQ is to the slope of the tangent line. If Δx is allowed to approach zero without reaching zero, then $(1 + 2\Delta x + \Delta x^2)$ will approach 1 without reaching 1. The slope of the tangent line at $(1,0)$ is 1.

c) **Equation of tangent line:**

$y - y_1 = m(x - x_1)$ Solve for y where $(x_1, y_1) = (1, 0)$

$y - 0 = 1(x - 1)$

$y - 0 = x - 1$

$y = x - 1$

13. Omitted.

17. Omitted.

21. Omitted.

Exercise 10.4

1. t = 1 second

   a) $d = 16t^2$ Calculate d = distance traveled

      $= 16(1)^2 = 16(1) = 16$ feet

   b) $s = 32t$ Calculate s = instantaneous speed

      $= 32(1) = 32$ feet/second

   c) Average speed $= \dfrac{\text{change in distance}}{\text{change in time}} = \dfrac{\Delta x}{\Delta t}$

      $= \dfrac{\text{distance at 1 second − distance at 0 seconds}}{1 \text{ second } - 0 \text{ seconds}}$

      $= \dfrac{16 - 0}{1 - 0} = \dfrac{16}{1} = 16$ feet/second

5. t = 8 seconds

   a) $d = 16t^2$ Calculate d = distance traveled

      $= 16(8)^2 = 16(64) = 1024$ feet

   b) $s = 32t$ Calculate s = instantaneous speed

      $= 32(8) = 256$ feet/second

   c) Average speed $= \dfrac{\text{change in distance}}{\text{change in time}} = \dfrac{\Delta x}{\Delta t}$

      $= \dfrac{\text{distance at 8 seconds − distance at 0 seconds}}{8 \text{ seconds } - 0 \text{ seconds}}$

      $= \dfrac{1024 - 0}{8 - 0} = \dfrac{1024}{8} = 128$ feet/second

9. $s = f(t) = 32t$, $f(t + \Delta t) = 32(t + \Delta t) = 32t + 32\Delta t$

$$f'(t) \text{ or } \frac{ds}{dt} = \lim_{\Delta t \to 0} \frac{f(t + \Delta t) - f(t)}{\Delta t} = \frac{\text{change in speed}}{\text{change in time}}$$

$$= \lim_{\Delta t \to 0} \frac{(32t + 32\Delta t) - 32t}{\Delta t}$$

$$= \lim_{\Delta t \to 0} \frac{32\Delta t}{\Delta t} = \lim_{\Delta t \to 0} 32 = 32$$

The "second fluxion of gravity" is the rate at which instantaneous speed is changing with respect to time. (This rate is the acceleration due to gravity.)

13. $f(x) = -7x + 42$,
     $f(x + \Delta x) = -7(x + \Delta x) + 42 = -7x - 7\Delta x + 42$

$$\frac{df}{dx} = \lim_{\Delta x \to 0} \frac{f(x + \Delta x) - f(x)}{\Delta x}$$

$$= \lim_{\Delta x \to 0} \frac{(-7x - 7\Delta x + 42) - (-7x + 42)}{\Delta x}$$

$$= \lim_{\Delta x \to 0} \frac{-7x - 7\Delta x + 42 + 7x - 42}{\Delta x} = \lim_{\Delta x \to 0} \frac{-7\Delta x}{\Delta x} = \lim_{\Delta x \to 0} -7 = -7$$

17. $f(x) = -7x + 42$, where $m = -7$, $b = 42$

$$\frac{df}{dx} = m = -7$$

21. $f(x) = 3x^2 - 11$,
     $f(x + \Delta x) = 3(x + \Delta x)^2 - 11$

$$= 3(x^2 + 2x(\Delta x) + \Delta x^2) - 11 = 3x^2 + 6x(\Delta x) + 3\Delta x^2 - 11$$

$$\frac{df}{dx} = \lim_{\Delta x \to 0} \frac{f(x + \Delta x) - f(x)}{\Delta x}$$

$$= \lim_{\Delta x \to 0} \frac{(3x^2 + 6x(\Delta x) + 3\Delta x^2 - 11) - (3x^2 - 11)}{\Delta x}$$

$$= \lim_{\Delta x \to 0} \frac{3x^2 + 6x(\Delta x) + 3\Delta x^2 - 11 - 3x^2 + 11}{\Delta x}$$

$$= \lim_{\Delta x \to 0} \frac{6x(\Delta x) + 3\Delta x^2}{\Delta x} = \lim_{\Delta x \to 0} \frac{\Delta x(6x + 3\Delta x)}{\Delta x}$$

$$= \lim_{\Delta x \to 0} (6x + 3\Delta x) = 6x$$

25. $f(x) = 11x^2$, where $a = 11$, $b = 0$, $c = 0$

$$\frac{df}{dx} = 2ax + b = 2(11)x + 0 = 22x$$

29. $f(x) = 3x^2 - 2x + 7$, where $a = 3$, $b = -2$, $c = 7$

$$\frac{df}{dx} = 2ax + b = 2(3)x + (-2) = 6x - 2$$

33. a) $f(x) = x^2 - 2x + 3$, where $a = 1$, $b = -2$, $c = 3$

$$\frac{df}{dx} = 2ax + b = 2(1)x + (-2) = 2x - 2$$

b)   Find the slope by substitution in df/dx.

At $x = 1$, $\frac{df}{dx} = 2x - 2 = 2(1) - 2 = 2 - 2 = 0$

At $x = 2$, $\frac{df}{dx} = 2x - 2 = 2(2) - 2 = 4 - 2 = 2$

c)   Find the tangent lines:

$m = 0$
At $x = 1$, $y = (1)^2 - 2(1) + 3 = 1 - 2 + 3 = 2$

$m = 2$
At $x = 2$, $y = (2)^2 - 2(2) + 3 = 4 - 4 + 3 = 3$

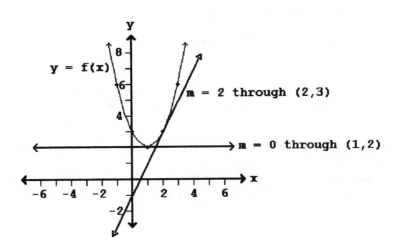

Exercise 10.4

37. $f(x) = 3x^3 + 4x - 1$

$f(x + \Delta x) = 3(x + \Delta x)^3 + 4(x + \Delta x) - 1$

$= 3(x^3 + 3x^2(\Delta x) + 3x(\Delta x)^2 + \Delta x^3) + 4x + 4\Delta x - 1$

$= 3x^3 + 9x^2(\Delta x) + 9x(\Delta x)^2 + 3\Delta x^3 + 4x + 4\Delta x - 1$

$\dfrac{df}{dx} = \lim_{\Delta x \to 0} \dfrac{f(x + \Delta x) - f(x)}{\Delta x}$

$= \lim_{\Delta x \to 0} \dfrac{3x^3 + 9x^2\Delta x + 9x(\Delta x)^2 + 3\Delta x^3 + 4x + 4\Delta x - 1 - 3x^3 - 4x + 1}{\Delta x}$

$= \lim_{\Delta x \to 0} \dfrac{9x^2(\Delta x) + 9x(\Delta x)^2 + 3\Delta x^3 + 4\Delta x}{\Delta x}$

$= \lim_{\Delta x \to 0} \dfrac{\Delta x(9x^2 + 9x\Delta x + 3\Delta x^2 + 4)}{\Delta x}$

$= \lim_{\Delta x \to 0} (9x^2 + 9x\Delta x + 3\Delta x^2 + 4) = 9x^2 + 4$

41. $f(x) = 2x^3 + x^2$, where $a = 2$, $b = 1$, $c = 0$, $d = 0$

$\dfrac{df}{dx} = 3ax^2 + 2bx + c = 3(2)x^2 + 2(1)x + 0 = 6x^2 + 2x$

45. $f(x) = 5x^4 - 3x^2 + 2x - 7$, where $a = 5$, $b = 0$, $c = -3$, $d = 2$, $e = -7$

$\dfrac{df}{dx} = 4ax^3 + 3bx^2 + 2cx + d$

$= 4(5)x^3 + 3(0)x^2 + 2(-3)x + 2$

$= 20x^3 - 6x + 2$

49. a) $\Delta f / \Delta x$ measures the average flow rate in gallons per hour.
    b) df/dx measures the instantaneous flow rate in gallons per hour.

53. a) $\Delta f / \Delta x$ measures the average growth rate of Metropolis in thousands of people per year.
    b) df/dx measures the instantaneous growth rate of Metropolis in thousands of people per year.

1. s = 250 ft/sec, c = 4 ft, angle of elevation = 30°

    a) $y = -\dfrac{64x^2}{3s^2} + \dfrac{x}{\sqrt{3}} + c$ 30° angle formula

         $= -\dfrac{64x^2}{3(250)^2} + \dfrac{x}{\sqrt{3}} + 4$ substituting

         $= -\dfrac{64x^2}{187,500} + 0.5774x + 4$

         $y = -0.0003x^2 + 0.5774x + 4$ rounding

    b) The distance that the ball travels is where y = 0.

    $0 = -0.0003x^2 + 0.5774x + 4$, a = -0.0003, b = 0.5774, c = 4

    $x = \dfrac{-b \pm \sqrt{b^2 - 4ac}}{2a}$

    $x = \dfrac{-0.5774 \pm \sqrt{(0.5774)^2 - 4(-0.0003)(4)}}{2(-0.0003)}$

    $= \dfrac{-0.5774 \pm \sqrt{0.3381908}}{-0.0006} = \dfrac{-0.5774 \pm 0.58154171}{-0.0006}$

    $x = \dfrac{-0.5774 + 0.58154}{-0.0006} = \dfrac{0.004142}{-0.0006} = -6.903 \approx -7$ feet

    or $x = \dfrac{-0.5774 - 0.58154}{-0.0006} = \dfrac{-1.15894}{-0.0006} = 1931.570 \approx 1932$ feet

    The negative value does not fit this situation, so the cannonball will travel 1932 feet.

    c) Find slope of the tangent line and let it equal zero.

    $m = \dfrac{df}{dx} = 2ax + b$ where a = -0.0003, b = 0.5774

         $= 2(-0.0003)x + 0.5774$

    $m = -0.0006x + 0.5774$ slope of the tangent line

1.c)   Continued

$0 = -0.0006x + 0.5774$ (solve for x with m = 0)

$0.0006x = 0.5774$

$x = \dfrac{0.5774}{0.0006} = 962.3333 \approx 962.3$ feet

Substitute x = 962.3 in equation of path for maximum height.

$y = -0.0003(962.3)^2 + 0.5774(962.3) + 4$

$= -277.81 + 555.63 + 4 = 281.82 \approx 282$ feet

d)   Substitute x = 1900 in equation of path to find the height of the cannonball.

$y = -0.0003(1900)^2 + 0.5774(1900) + 4$

$= -1083.00 + 1,097.06 + 4 = 18.06 \approx 18$ feet

Since the height of the cannonball at 1900 feet from the cannon is 18 feet, the cannonball would not clear the 80-foot wall.

e)

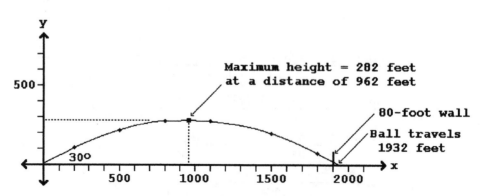

5.   s = 325 ft/sec, c = 5 ft, angle of elevation = 60°

a) $y = -\dfrac{64x^2}{s^2} + \sqrt{3}x + c$ 60° angle formula

$= -\dfrac{64x^2}{(325)^2} + \sqrt{3}x + 5$ substituting

$= -\dfrac{64x^2}{105,625} + 1.7321x + 5$

$y = -0.0006x^2 + 1.7321x + 5$ rounding

5.   Continued

b)   The distance that the ball travels is where y = 0.

0 = -0.0006x² + 1.7321x + 5, a = -0.0006, b = 1.7321, c = 5

$$x = \frac{-b \pm \sqrt{b^2 - 4ac}}{2a}$$

$$x = \frac{-1.7321 \pm \sqrt{(1.7321)^2 - 4(-0.0006)(5)}}{2(-0.0006)}$$

$$= \frac{-1.7321 \pm \sqrt{3.0121704}}{-0.0012} = \frac{-1.7321 \pm 1.735561}{-0.0012}$$

$$x = \frac{-1.7321 + 1.7356}{-0.0012} = \frac{0.0035}{-0.0012} = -2.9 \approx -3 \text{ feet}$$

$$\text{or } x = \frac{-1.7321 - 1.7356}{-0.0012} = \frac{-3.4677}{-0.0012} = 2889.75 \approx 2890 \text{ feet}$$

The negative value does not fit this situation, so the cannonball will travel 2890 feet.

c)   Find slope of the tangent line and let it equal zero.

$$m = \frac{df}{dx} = 2ax + b \quad \text{where a = -0.0006, b = 1.7321}$$

$$= 2(-0.0006)x + 1.7321$$

m = -0.0012x + 1.7321   slope of the tangent line

0 = -0.0012x + 1.7321   (solve for x with m = 0)

0.0012x = 1.7321

$$x = \frac{1.7321}{0.0012} = 1443.4167 \approx 1443 \text{ feet}$$

Substitute x = 1443 in equation of path for maximum height.

y = -0.0006(1443)² + 1.7321(1443) + 5

= -1249.35 + 2499.42 + 5 = 1255.07 ≈ 1255 feet

d)   Substitute x = 1900 in equation of path to find the height of the cannonball.

y = -0.0006(1900)² + 1.7321(1900) + 5

= -2166 + 3290.99 + 5 = 1129.99 ≈ 1130 feet

Since the height of the cannonball at 1900 feet from the cannon is 1130 feet, the cannonball would clear the 80-foot wall.

5.   Continued

    e)

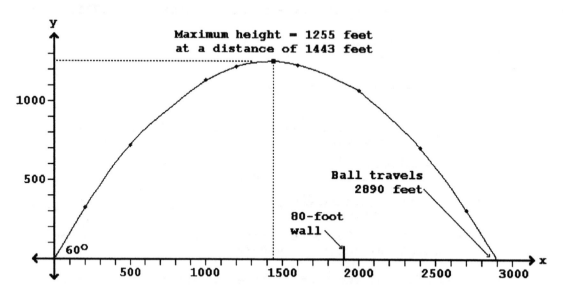

9.   s = 250 ft/sec, c = 4 ft, angle of elevation = 45°

    a) $y = -\dfrac{32x^2}{s^2} + x + c$

         $= -\dfrac{32x^2}{(250)^2} + x + 4$

         $= -\dfrac{32x^2}{62,500} + x + 4$

         $y = -0.0005x^2 + x + 4$

    b)   The distance that the ball travels is where y = 0.

         $0 = -0.0005x^2 + x + 4$, a = -0.0005, b = 1, c = 4

         $x = \dfrac{-b \pm \sqrt{b^2 - 4ac}}{2a}$

         $x = \dfrac{-1 \pm \sqrt{(1)^2 - 4(-0.0005)(4)}}{2(-0.0005)}$

         $= \dfrac{-1 \pm \sqrt{1.008}}{-0.0010} = \dfrac{-1 \pm 1.003992}{-0.0010}$

9.b) Continued

$$x = \frac{-1 + 1.003992}{-0.0010} = \frac{0.003992}{-0.0010} = -3.99 \approx -4 \text{ feet}$$

$$\text{or } x = \frac{-1 - 1.003992}{-0.0010} = \frac{-2.003992}{-0.0010} = 2003.99 \approx 2004 \text{ feet}$$

The negative value does not fit this situation, so the cannonball will travel 2004 feet.

c) Find slope of the tangent line and let it equal zero.

$$m = \frac{df}{dx} = 2ax + b \qquad \text{where } a = -0.0005, \ b = 1$$

$$= 2(-0.0005)x + 1$$

$$m = -0.0010x + 1 \qquad \text{slope of the tangent line}$$

$$0 = -0.0010x + 1 \qquad (\text{solve for } x \text{ with } m = 0)$$

$$0.0010x = 1$$

$$x = \frac{1}{0.0010} = 1000 \text{ feet}$$

Substitute x = 1000 in equation of path for maximum height.

$$y = -0.0005(1000)^2 + 1000 + 4$$

$$= -500 + 1000 + 4 = 504 \text{ feet}$$

d) Substitute x = 1100 in equation of path to find the height of the cannonball.

$$y = -0.0005(1100)^2 + 1100 + 4$$

$$= -605 + 1100 + 4 = 499 \text{ feet}$$

Since the height of the cannonball at 1100 feet from the cannon is 499 feet, the cannonball would clear the 70-foot wall.

9.   Continued

   e)

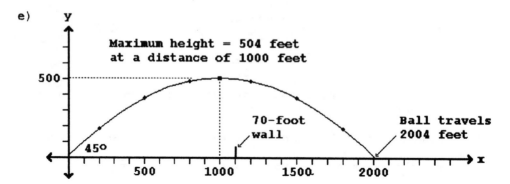

13.   s = 240 ft/sec, c = 4 ft, angle of elevation = 45°

   a) $y = -\dfrac{32x^2}{s^2} + x + c$

   $= -\dfrac{32x^2}{(240)^2} + x + 4$

   $= -\dfrac{32x^2}{57,600} + x + 4$

   $y = -0.0006x^2 + x + 4$

   b)   Find x when y = 20

   $20 = -0.0006x^2 + x + 4$

   $0 = -0.0006x^2 + x - 16$,   a = -0.0006, b = 1, c = -16

   $x = \dfrac{-b \pm \sqrt{b^2 - 4ac}}{2a}$

   $x = \dfrac{-1 \pm \sqrt{(1)^2 - 4(-0.0006)(-16)}}{2(-0.0006)}$

   $= \dfrac{-1 \pm \sqrt{0.9616}}{-0.0012} = \dfrac{-1 \pm 0.9806}{-0.0012}$

   $x = \dfrac{-1 + 0.9806}{-0.0012} = \dfrac{-0.0194}{-0.0012} = 16.167 \approx 16$ feet

   or $x = \dfrac{-1 - 0.9806}{-0.0012} = \dfrac{-1.9806}{-0.0012} = 1650.5$ feet

   The cannon should be placed 1651 feet from the target in
   order for the cannonball to hit it and keep the crew as far
   as possible from the enemy.

13. Continued

c)

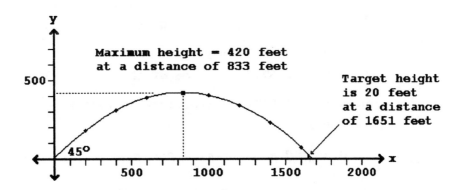

Maximum height = 420 feet
at a distance of 833 feet

Target height
is 20 feet
at a distance
of 1651 feet

$45°$

17. a) c = 3 ft, angle of elevation = 30°, x = 400, y = 10. Find s.

$$y = -\frac{64x^2}{3s^2} + \frac{x}{\sqrt{3}} + c \qquad \text{30° angle formula}$$

$$10 = -\frac{64(400)^2}{3s^2} + \frac{400}{\sqrt{3}} + 3 \qquad \text{substituting}$$

$$10 = -\frac{3,413,333}{s^2} + 230.9401 + 3$$

$$10 - 233.940 = -\frac{3,413,333}{s^2}$$

$$-223.940s^2 = -3,413,333$$

$$s^2 = \frac{-3,413,333}{-223.940} = 15,242.170$$

$$s = 123.459 \approx 123 \text{ ft/sec}$$

b) $$y = -\frac{64x^2}{3s^2} + \frac{x}{\sqrt{3}} + c \qquad \text{30° angle formula}$$

$$= -\frac{64x^2}{3(123)^2} + \frac{x}{\sqrt{3}} + 3 \qquad \text{substituting}$$

$$= -\frac{64x^2}{45,387} + 0.5774x + 3$$

$$y = -0.0014x^2 + 0.5774x + 3 \qquad \text{rounding}$$

Exercise 10.5

17. Continued

   c) Find slope of the tangent line and let it equal zero.

   $m = \dfrac{df}{dx} = 2ax + b$ where $a = -0.0014$, $b = 0.5774$

   $= 2(-0.0014)x + 0.5774$

   $m = -0.0028x + 0.5774$ slope of the tangent line

   $0 = -0.0028x + 0.5774$ (solve for x with m = 0)

   $0.0028x = 0.5774$

   $x = \dfrac{0.5774}{0.0028} = 206.2143 \approx 206$ feet

   Substitute x = 206 in equation of path for maximum height.

   $y = -0.0014(206)^2 + 0.5774(206) + 3$

   $= -59.41 + 118.94 + 3 = 62.53 \approx 63$ feet

   d) c = 3 ft, angle of elevation = 45°, x = 400, y = 10.   Find s.

   $y = -\dfrac{32x^2}{s^2} + x + c$ 45° angle formula

   $10 = -\dfrac{32(400)^2}{s^2} + 400 + 3$

   $10 = -\dfrac{5,120,000}{s^2} + 400 + 3$

   $10 - 403 = -\dfrac{5,120,000}{s^2}$

   $-393s^2 = -5,120,000$

   $s^2 = \dfrac{-5,120,000}{-393} = 13027.9898$

   $s = 114.14 \approx 114$ ft/sec

21. a) Since there are 180° in a triangle, the remaining angle is:

   $x = 180° - (30° + 90°)$

   $= 180° - 120°$

   $= 60°$

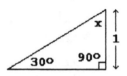

21. **Continued**

b) All three angles are 60° (see drawing).

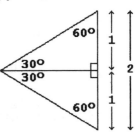

c) The length of the vertical side = 1 + 1 = 2.

d) Since all three angles are 60°, the triangle is an equilateral triangle and all three sides are equal. Each side of the larger triangle would equal 2.

e) a = 1, b = ?, c = 2

$$a^2 + b^2 = c^2$$
$$1^2 + b^2 = 2^2$$
$$1 + b^2 = 4$$
$$b^2 = 3$$
$$b = \sqrt{3}$$

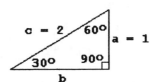

25. s = 200 ft/sec, c = 5 ft, angle of elevation = 20°

a) $y = -16\dfrac{x^2}{(s\ \cos\theta)^2} + x\ \tan\theta + c$ general angle formula

$y = -16\dfrac{x^2}{(200\ \cos\ 20°)^2} + x\ \tan\ 20° + 5$ substituting

$= -16\dfrac{64x^2}{35320.88886} + 0.3639702343x + 5$

$y = -0.0004529897x^2 + 0.3639702343x + 5$

$y = -0.0005x^2 + 0.3640x + 5$ rounding

b) The distance that the ball travels is where y = 0.

$0 = -0.0005x^2 + 0.3640x + 5$, a = -0.0005, b = 0.3640, c = 5

$x = \dfrac{-b \pm \sqrt{b^2 - 4ac}}{2a}$

$x = \dfrac{-0.3640 \pm \sqrt{(0.3640)^2 - 4(-0.0005)(5)}}{2(-0.0005)}$

25.b) Continued

$$= \frac{-0.3640 \pm \sqrt{0.142496}}{-0.0010} = \frac{-0.3640 \pm 0.377486}{-0.0010}$$

$$x = \frac{-0.3640 + 0.377486}{-0.0010} = \frac{0.013486}{-0.0010} = -13.486 \approx -13 \text{ feet}$$

$$\text{or } x = \frac{-0.3640 - 0.377486}{-0.0010} = \frac{-0.741486}{-0.0010} = 741.486 \approx 741 \text{ feet}$$

The negative value does not fit this situation, so the cannonball will travel 741 feet.

c)   Find slope of the tangent line and let it equal zero.

$$m = \frac{df}{dx} = 2ax + b \qquad \text{where } a = -0.0005, \ b = 0.3640$$

$$= 2(-0.0005)x + 0.3640$$

$$m = -0.0010x + 0.3640 \qquad \text{slope of the tangent line}$$

$$0 = -0.0010x + 0.3640 \qquad (\text{solve for } x \text{ with } m = 0)$$

$$0.0010x = 0.3640$$

$$x = \frac{0.3640}{0.0010} = 364 \text{ feet}$$

Substitute x = 364 in equation of path for maximum height.

$$y = -0.0005(364)^2 + 0.3640(364) + 5$$

$$= -66.248 + 132.496 + 5 = 71.248 \approx 71 \text{ feet}$$

d)

Exercise 10.6

1. Find the function whose derivative is $f(x) = 5 = 5x^0$.

    FIRST GUESS (power): Increase the power by 1. Need x^1 because the derivative of x^1 is 1. However, we want the derivative to be 5.

    SECOND GUESS (coefficient): Place a 5 in front of the x to make 5x. The derivative of 5x is 5.

    ANSWER: The antiderivative of $f(x) = 5$ is $5x + c$

    or $\quad \int 5dx = 5x + c$

5. The region is shown in the following graph

    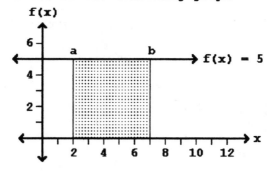

    a) $A(x) = \int 5dx = 5x + c \qquad$ from Exercise 1.

        To find the area of a region find
            A(right boundary) - A(left boundary)

        left boundary = a = 2, right boundary = b = 7

        $$A(b) - A(a) = A(7) - A(2)$$
        $$= [5(7) + c] - [5(2) + c]$$
        $$= (35 + c) - (10 + c)$$
        $$= 35 + c - 10 - c$$
        $$= 25$$

    b) The region is a rectangle with base = 5 and height = 5.

        $A = bh = 5(5) = 25$

Exercise 10.6

9.   Find the function whose derivative is

$$g(x) = 8x + 7 = 8x^1 + 7x^0$$

FIRST GUESS (powers): Increase the powers by 1. Need $x^2 + x^1$ because the derivative of $x^2 + x^1$ is $2x + 1$. However, we want the derivative to be $8x + 7$.

SECOND GUESS (coefficients): Place a 4 in front of the x^2 and a 7 in front of the x to make $4x^2 + 7x$. The derivative of $4x^2 + 7x$ is $8x + 7$.

ANSWER: The antiderivative of $g(x) = 8x + 7$ is $4x^2 + 7x + c$

or $\int (8x + 7) dx = 4x^2 + 7x + c$

13. The region is shown in the following graph

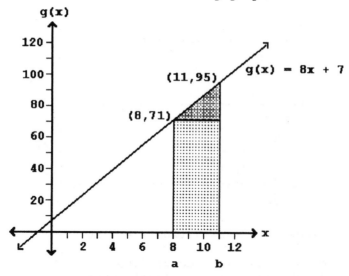

$A(x) = \int (8x + 7) dx = 4x^2 + 7x + c$ from Exercise 9.

To find the area of a region calculate
     A(right boundary) − A(left boundary)

left boundary = a = 8, right boundary = b = 11

$$
\begin{aligned}
A(b) - A(a) &= A(11) - A(8) \\
&= [4(11)^2 + 7(11) + c] - [4(8)^2 + 7(8) + c] \\
&= (484 + 77 + c) - (256 + 56 + c) \\
&= 561 + c - 312 - c \\
&= 249
\end{aligned}
$$

13. Continued

Check the solution:

The region is a rectangle with base = 3 and height = 71 plus a triangle of base = 3 and height = 95 – 71 = 24.

$$A = A(\text{rectangle}) + A(\text{triangle}) = bh + \frac{1}{2}bh$$

$$= 3(71) + \frac{1}{2}(3)(24) = 213 + 36 = 249$$

17. a) $f(x) = x$, where $m = 1$, $b = 0$

$$\frac{df}{dx} = m = 1$$

b)   Find the function whose derivative is $f(x) = x = 1x^1$.

FIRST GUESS (power): Increase the power by 1. Need x^2 because the derivative of x^2 is $2x$. However, we want the derivative to be x.

SECOND GUESS (coefficient): Place a 1/2 in front of the x^2 to make $(1/2)x^2$. The derivative of $(1/2)x^2$ is x.

ANSWER: The antiderivative of $f(x) = x$ is $(1/2)x^2 + c$

or $\int x\,dx = \frac{1}{2}x^2 + c$

21. a) $m(x) = 8x$, where $a = 8$, $b = 0$

$$\frac{dm}{dx} = a = 8$$

b)   Find the function whose derivative is $m(x) = 8x = 8x^1$

FIRST GUESS (powers): Increase the powers by 1. Need x^2 because the derivative of x^2 is $2x$. However, we want the derivative to be $8x$.

SECOND GUESS (coefficients): Place a 4 in front of the x^2 to make $4x^2$. The derivative of $4x^2$ is $8x$.

ANSWER: The antiderivative of $m(x) = 8x$ is $4x^2 + c$

or $\int 8x^2\,dx = 4x^2 + c$

25.   Find the function whose derivative is $f(x) = 3x^2$

FIRST GUESS (powers):   Increase the powers by 1.   Need x^3 because the derivative of x^3 is $3x^2$.

SECOND GUESS (coefficients):   Not necessary.

ANSWER: The antiderivative of $f(x) = 3x^2$ is $x^3 + c$

or $A(x) = \int 3x^2 dx = x^3 + c$

To find the area of a region calculate
A(right boundary) − A(left boundary)

a)   left boundary = a = 1, right boundary = b = 4

$$
\begin{aligned}
A(b) - A(a) &= A(4) - A(1) \\
&= [(4)^3 + c] - [(1)^3 + c] \\
&= (64 + c) - (1 + c) \\
&= 64 + c - 1 - c \\
&= 63
\end{aligned}
$$

b)   left boundary = a = 3, right boundary = b = 100

$$
\begin{aligned}
A(b) - A(a) &= A(100) - A(3) \\
&= [(100)^3 + c] - [(3)^3 + c] \\
&= (1{,}000{,}000 + c) - (27 + c) \\
&= 1{,}000{,}000 + c - 27 - c \\
&= 999{,}973
\end{aligned}
$$

29.   Find the function whose derivative is

$$f(x) = 5x^6 + 3 = 5x^6 + 3x^0$$

FIRST GUESS (powers):   Increase the powers by 1.   Need $x^7 + x^1$ because the derivative of $x^7 + x^1$ is $7x^6 + 1$.   However, we want the derivative to be $5x^6 + 3$.

SECOND GUESS (coefficients):   Place 5/7 in front of the x^7 and a 3 in front of the x to make $\frac{5}{7}x^7 + 3x$.

ANSWER: $A(x) = \int (5x^6 + 3)\, dx = \frac{5}{7}x^7 + 3x + c$

To find the area of a region calculate
A(right boundary) − A(left boundary)

29. Continued

a) left boundary = a = 4, right boundary = b = 7

$$A(b) - A(a) = A(7) - A(4)$$

$$= \left(\frac{5}{7}(7)^7 + 3(7) + c\right) - \left(\frac{5}{7}(4)^7 + 3(4) + c\right)$$

$$= (588,245 + 21 + c) - (11,702.86 + 12 + c)$$
$$= 588,266 + c - 11,714.86 - c$$
$$= 576,551.14$$

b) left boundary = a = 3, right boundary = b = 10

$$A(b) - A(a) = A(10) - A(3)$$

$$= \left(\frac{5}{7}(10)^7 + 3(10) + c\right) - \left(\frac{5}{7}(3)^7 + 3(3) + c\right)$$

$$= (7,142,857 + 30 + c) - (1,562 + 9 + c)$$
$$= 7,142,887 + c - 1,571 - c$$
$$= 7,141,316$$

33. $g(x) = 8x + 7 = 8x^1 + 7x^0$ Break into two separate problems added together

$$\int ax^n dx = \frac{a}{n+1} x^{n+1} + c$$ where $a_1 = 8$, $n_1 = 1$ and $a_2 = 7$, $n_2 = 0$

$$\int (8x^1 + 7x^0)\,dx = \int 8x^1 dx + \int 7x^0 dx$$

$$= \frac{8}{1+1}x^{1+1} + \frac{7}{0+1}x^{0+1} + c$$

$$= \frac{8}{2}x^2 + \frac{7}{1}x^1 + c = 4x^2 + 7x + c$$

37. $f(x) = 5x^6 - 3x^2 + 13$ Break into three separate problems added together

$$\int ax^n dx = \frac{a}{n+1} x^{n+1} + c$$ where $a_1 = 5$, $n_1 = 6$, $a_2 = -3$, $n_2 = 2$, $a_3 = 13$, $n_3 = 0$

$$\int (5x^6 - 3x^2 + 13)\,dx = \int 5x^6 dx - \int 3x^2 dx + \int 13x^0 dx$$

$$= \frac{5}{6+1}x^{6+1} - \frac{3}{2+1}x^{2+1} + \frac{13}{0+1}x^{0+1} + c$$

$$= \frac{5}{7}x^7 - \frac{3}{3}x^3 + \frac{13}{1}x^1 + c = \frac{5}{7}x^7 - x^3 + 13x + c$$

Chapter 10 Review

1.   Omitted.

5.   Time in motion until time B = AB = 30 seconds,
     Speed at time B = BC = 8 feet per second

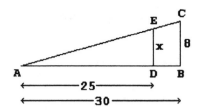

a)   distance = area of triangle ABC = $\frac{1}{2}$bh

     d = $\frac{1}{2}$(30)(8) = 120 feet

b)   time in motion until time D = AD = 25 seconds
     Speed at time D = DE = x

     $\dfrac{\text{speed at time B}}{\text{speed at time D}} = \dfrac{\text{time in motion until time B}}{\text{time in motion until time D}}$

     $\dfrac{8}{x} = \dfrac{30}{25}$

     30x = 8(25) = 200

     x = $\dfrac{200}{30}$ = 6.6667 ≈ 6.7 feet per second

c)   distance = area of triangle ADE = $\frac{1}{2}$bh

     d = $\frac{1}{2}$(25)$\left(\dfrac{20}{3}\right)$ = $\dfrac{500}{6}$ = 83.3333 ≈ 83.3 feet

d)   average speed = $\dfrac{\text{distance traveled in 25 seconds}}{25 \text{ seconds}}$

     average speed = $\dfrac{83.3}{25}$ = 3.333 ≈ 3.3 feet per second

9. $y = x^2 + 4x + 11$

a)   Substituting for x yields the following points:

x	$y = x^2 + 4x + 11$	y	(x,y)
0	$(0)^2 + 4(0) + 11 = 0 + 0 + 11$	11	(0,11)
1	$(1)^2 + 4(1) + 11 = 1 + 4 + 11$	16	(1,16)
2	$(2)^2 + 4(2) + 11 = 4 + 8 + 11$	23	(2,23)

Look at the slope of the secant lines to determine which direction the vertex lies:

Points (0,11) and (1,16): $m = \dfrac{\Delta y}{\Delta x} = \dfrac{16 - 11}{1 - 0} = 5$

Points (1,16) and (2,23): $m = \dfrac{\Delta y}{\Delta x} = \dfrac{23 - 16}{2 - 1} = 7$

The vertex lies to the left of x = 0 since 5 is closer to zero than 7.

x	$y = x^2 + 4x + 11$	y	(x,y)
-1	$(-1)^2 + 4(-1) + 11 = 1 - 4 + 11$	8	(-1,8)
-2	$(-2)^2 + 4(-2) + 11 = 4 - 8 + 11$	7	(-2,7)
-3	$(-3)^2 + 4(-3) + 11 = 9 - 12 + 11$	8	(-3,8)
-4	$(-4)^2 + 4(-4) + 11 = 16 - 16 + 11$	11	(-4,11)

9.a)   Continued

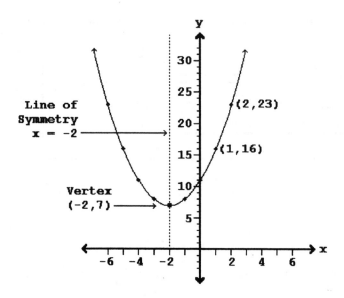

b)   The y-values are the same at x = -3 and x = -1, so the
     equation of the line of symmetry is x = -2.

c)   The vertex is the point (-2,7).

d)   Using points (1,16) and (2,23), the slope of the secant
     line is

$$m = \frac{\Delta y}{\Delta x} = \frac{23 - 16}{2 - 1} = \frac{7}{1} = 7$$

e)   **Cauchy's method:**

P is at $x_1 = 1$, $y_1 = 16$
Q is at $x_2 = 1 + \Delta x$, $y_2 = (1 + \Delta x)^2 + 4(1 + \Delta x) + 11$

$$= (1 + 2\Delta x + \Delta x^2) + 4 + 4\Delta x + 11$$

$$= 16 + 6\Delta x + \Delta x^2$$

slope of PQ $= \dfrac{\Delta y}{\Delta x} = \dfrac{y_2 - y_1}{x_2 - x_1}$ where PQ is the secant line

$$= \frac{(16 + 6\Delta x + \Delta x^2) - 16}{(1 + \Delta x) - 1}$$

$$= \frac{6\Delta x + \Delta x^2}{\Delta x} = \frac{\Delta x(6 + \Delta x)}{\Delta x} = 6 + \Delta x$$

9.e) Continued

The closer Δx is to 0, the closer Q is to P, and the closer the slope of the secant line PQ is to the slope of the tangent line. If Δx is allowed to approach zero without reaching zero, then $6 + \Delta x$ will approach 6 without reaching 6. The slope of the tangent line at (1,16) is 6.

f)   <u>Equation of tangent line:</u>

$y - y_1 = m(x - x_1)$ Solve for y where $(x_1, y_1) = (1, 16)$

$y - 16 = 6(x - 1)$
$y - 16 = 6x - 6$
$\qquad y = 6x + 10$

g)

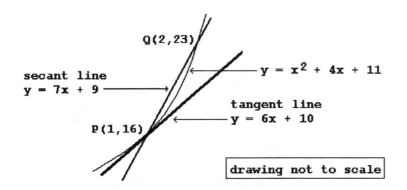

13. $f(x) = 2x - 15$
$f(x + \Delta x) = 2(x + \Delta x) - 15 = 2x + 2\Delta x - 15$

$\dfrac{df}{dx} = \lim_{\Delta x \to 0} \dfrac{f(x + \Delta x) - f(x)}{\Delta x}$

$\qquad = \lim_{\Delta x \to 0} \dfrac{(2x + 2\Delta x - 15) - (2x - 15)}{\Delta x}$

$\qquad = \lim_{\Delta x \to 0} \dfrac{2x + 2\Delta x - 15 - 2x + 15}{\Delta x}$

$\qquad = \lim_{\Delta x \to 0} \dfrac{2\Delta x}{\Delta x} = \lim_{\Delta x \to 0} 2 = 2$

17. s = 200 ft/sec, c = 4.5 ft, angle of elevation = 30°

a) $y = -\dfrac{64x^2}{3s^2} + \dfrac{x}{\sqrt{3}} + c$ 30° angle formula

   $= -\dfrac{64x^2}{3(200)^2} + \dfrac{x}{\sqrt{3}} + 4.5$ substituting

   $= -\dfrac{64x^2}{120,000} + 0.5774x + 4.5$

   $y = -0.0005x^2 + 0.5774x + 4.5$ rounding

b) The distance that the ball travels is where y = 0.

   $0 = -0.0005x^2 + 0.5774x + 4.5$, a = -0.0005, b = 0.5774, c = 4.5

   $x = \dfrac{-b \pm \sqrt{b^2 - 4ac}}{2a}$

   $x = \dfrac{-0.5774 \pm \sqrt{(0.5774)^2 - 4(-0.0005)(4.5)}}{2(-0.0005)}$

   $= \dfrac{-0.5774 \pm \sqrt{0.3423908}}{-0.001} = \dfrac{-0.5774 \pm 0.585142}{-0.001}$

   $x = \dfrac{-0.5774 + 0.585142}{-0.001} = \dfrac{0.007742}{-0.001} = -7.7$ feet

   or $x = \dfrac{-0.5774 - 0.585142}{-0.001} = \dfrac{-1.1625}{-0.001} = 1162.5$ feet

   The negative value does not fit this situation, so the cannonball will travel 1163 feet.

c) Find slope of the tangent line and let it equal zero.

   $m = \dfrac{df}{dx} = 2ax + b$ where a = -0.0005, b = 0.5774

   $= 2(-0.0005)x + 0.5774$

   $m = -0.001x + 0.5774$ slope of the tangent line

   $0 = -0.001x + 0.5774$ (solve for x with m = 0)

   $0.001x = 0.5774$

   $x = \dfrac{0.5774}{0.001} = 577.4$ feet

17.c) Continued

Substitute x = 577.4 in equation of path for maximum height

y = -0.0005(577.4) + 0.5774(577.4) + 4.5

= -166.70 + 333.39 + 4.5 = 171.19 ≈ 171 feet

d) Substitute x = 100 in equation of path for height

y = -0.0005(100) + 0.5774(100) + 4.5

= -5 + 57.74 + 4.5 = 57.24 ≈ 57 feet

Since the height at 100 feet from the cannon is 57 feet, the cannonball will clear the 45-foot wall.

e)

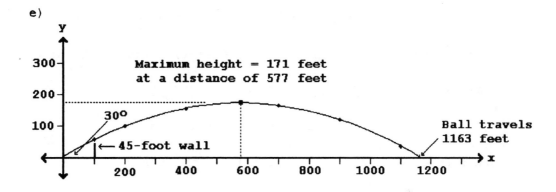

21. Omitted.

25. Omitted.

29. Omitted.

Appendix V   Dimensional Analysis

Exercises

1. Standard Conversion: 1 yard = 3 feet

    a) $12 \text{ feet} = 12 \text{ feet}\left(\dfrac{1 \text{ yard}}{3 \text{ feet}}\right) = 4 \text{ yards}$

    b) $12 \text{ yards} = 12 \text{ yards}\left(\dfrac{3 \text{ feet}}{1 \text{ yard}}\right) = 12(3) \text{ feet} = 36 \text{ feet}$

5. Standard Conversions: 1 mile = 5280 feet, 1 foot = 12 inches

    $2 \text{ miles} = \left(\dfrac{5280 \text{ feet}}{1 \text{ mile}}\right) = 2(5280 \text{ feet})\left(\dfrac{12 \text{ inches}}{1 \text{ foot}}\right)$

    $= 2(5280)(12) \text{ inches} = 126,720 \text{ inches}$

9. Standard Conversion: $1 \text{ centiliter} = \dfrac{1}{100} \text{ liter}$

    a) $2 \text{ centiliters} = 2 \text{ centiliters}\left(\dfrac{1 \text{ liter}}{100 \text{ centiliters}}\right)$

    $= \dfrac{2(1) \text{ liters}}{100} = 0.02 \text{ liters}$

    b) $2 \text{ liters} = 2 \text{ liters}\left(\dfrac{100 \text{ centiliters}}{1 \text{ liters}}\right)$

    $= 2(100 \text{ centiliters}) = 200 \text{ centiliters}$

13. Standard Conversions: 1 mile = 5280 feet, 1 hour = 60 minutes
                            1 minute = 60 seconds

    a) $\dfrac{60 \text{ miles}}{\text{hour}} = \dfrac{60 \text{ miles}}{\text{hour}}\left(\dfrac{5280 \text{ feet}}{1 \text{ miles}}\right) = \dfrac{60(5280) \text{ feet}}{\text{hour}}$

    $\dfrac{60(5280) \text{ feet}}{\text{hour}} = \dfrac{60(5280) \text{ feet}}{\text{hour}}\left(\dfrac{1 \text{ hour}}{60 \text{ minutes}}\right) = 5280 \text{ feet/minute}$

13.a)   Continued

$$\frac{5280 \text{ feet}}{\text{minute}} = \frac{5280 \text{ feet}}{\text{minute}}\left(\frac{1 \text{ minute}}{60 \text{ seconds}}\right) = \frac{5280 \text{ feet}}{60 \text{ seconds}}$$

$$= 88 \text{ feet/second}$$

b)   Distance = 80 feet, rate = 88 feet/second, find time.

$$d = rt \text{ or } t = \frac{d}{r}$$

$$t = \frac{80 \text{ feet}}{\left(\frac{88 \text{ feet}}{\text{second}}\right)} = 80 \text{ feet}\left(\frac{1 \text{ second}}{88 \text{ feet}}\right) = \frac{80}{88} \text{ seconds}$$

$$= 0.90909\ldots \text{ seconds} \approx 0.9 \text{ seconds}$$

17.   Standard Conversion: 1 year = 365 days

a) $\dfrac{5.18\%}{1 \text{ year}} = \dfrac{5.18\%}{1 \text{ year}}\left(\dfrac{1 \text{ year}}{365 \text{ days}}\right) = \dfrac{5.18\%}{365 \text{ days}} = 0.0141917808\%/\text{day}$

b)   Principal = $10,000

Interest = (rate/day)(principal)(days/month)

$$= \left(\frac{0.0141917808\%}{\text{day}}\right)(\$10,000)\left(\frac{30 \text{ days}}{\text{month of September}}\right)$$

$$= \frac{(0.000141917808)(\$10,000)(30)}{\text{month of September}}$$

$$= 42.5753425 \approx \$42.57 \text{ for September}$$

c)   Principal = $10,000

Interest = (rate/day)(principal)(days/month)

$$= \left(\frac{0.0141917808\%}{\text{day}}\right)(\$10,000)\left(\frac{31 \text{ days}}{\text{month of October}}\right)$$

$$= \frac{(0.000141917808)(\$10,000)(31)}{\text{month of October}}$$

$$= 43.9945205 \approx \$43.99 \text{ for October}$$